Oxford
comprehensive
mathematics

A secondary course for mixed abilities

Book 2
D. Paling   M. E. Wardle
illustrated by Gunvor Edwards

Oxford University Press 1975

Oxford University Press, Ely House, London W1

Glasgow New York Toronto Melbourne Wellington
Cape Town Ibadan Nairobi Dar es Salaam Lusaka Addis Ababa
Delhi Bombay Calcutta Madras Karachi Lahore Dacca
Kuala Lumpur Singapore Hong Kong Tokyo

This course was devised by:

C. S. Banwell
J. E. Hiscocks
D. Paling
K. D. Saunders
M. E. Wardle
C. J. Weeks

The course consists of:

Books 1 and 2
     for mixed ability classes
Books 3, 4 and 5
     (blue covers) for GCE
Books 3, 4 and 5
     (green covers) for CSE

Each book is accompanied by a teacher's book.
Books 1, 2 and Books 3 and 4 for CSE are accompanied by workbooks. The abbreviation (W/s) in the text signifies there is a worksheet for this material in the workbook.

In this book:

▌ indicates second level material

▌ indicates third level material

We gratefully acknowledge the permission of John Murray Ltd. and J. B. Lippincott Co. to use Alfred Noyes's 'The Highwayman' from *Collected Poems*. Copyright 1913, renewed 1941 by Alfred Noyes.

Filmset in Monophoto Times
by BAS Printers Limited, Wallop, Hampshire
and printed in Great Britain by
Hazell Watson & Viney Ltd, Aylesbury, Bucks

Contents

Flow charts

Think of a number between one and twenty	Write down a number with three different digits (e.g. 731)
Double it	Reverse the order of the digits (137)
Add 6	Subtract the smaller from the larger (731 − 137)
Divide by 2	Reverse the order of the digits of your last result
Subtract 3	Add this new number to the result of the subtraction
Write down the answer	Write down the answer
What do you notice?	

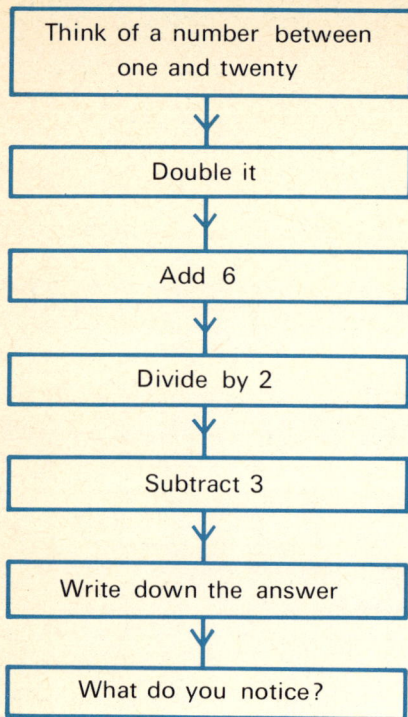

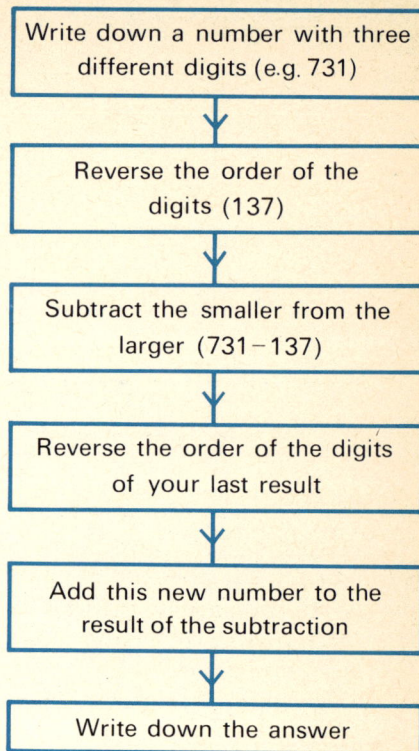

A   Try carrying out the instructions in the flow charts above.
Make a note of the number you start with and the number you finish with in each case.

B   What do you notice about the number you finish with in the left-hand flow chart?
Try again with some other numbers.

C   Compare your answer for the right-hand flow chart with those of your friends. Are they the same?
Now try again using a different three-digit number.
Compare your answers again. What do you notice?

D   Can you explain why, in the first flow chart, the number you finish with is the same as the number you start with?

Think of a number between 1 and 10
↓
Multiply by 5
↓
Add 2
↓
Double it
↓
Subtract 3
↓
Write down the answer

Think of a number between 1 and 10
↓
Double it
↓
Add 1
↓
Multiply by 5
↓
Subtract 4
↓
Write down the answer

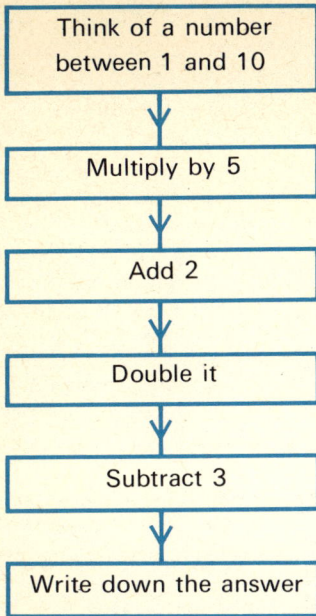

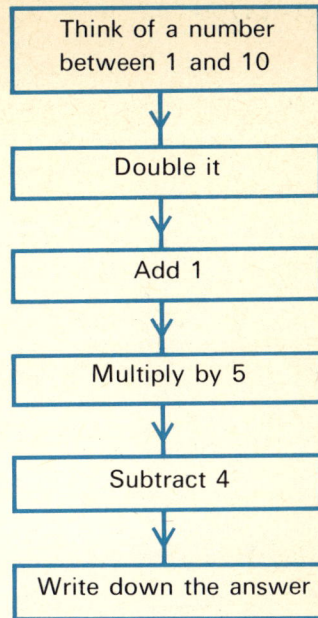

A Carry out the instructions in the flow charts above.
Use the same number to start with in both.
What do you notice about your two answers?

B Try using other numbers in each flow chart.
Copy the arrow graph on the right, and use your results to complete it.

Numbers thought of
2
3
4
5
6
7
8
9

Answers using flow chart
71

C Could you complete the arrow graph for the number 10 without using the flow chart? Could you complete the arrow graph for
1. 20?   2. 35?

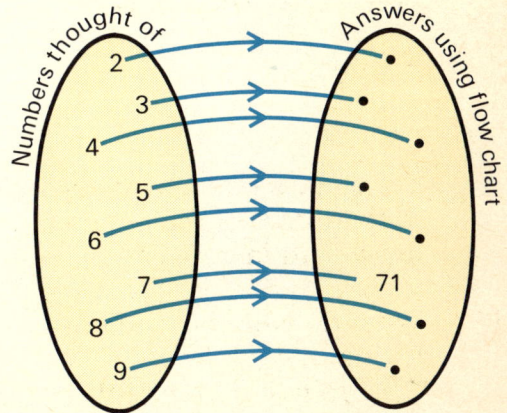

D Write down in your own words how you obtained the answer for any of the numbers in the arrow graph.

E Can you explain why the two flow charts give the same answer if you start with the same number?

F Draw a simpler version of the flow charts to show your answer to D.

START

Look at the cinema advertisements in the newspaper

Is there a film which you would like to see ? — No

Yes

Is it an X or an AA ? — Yes → Think of something else to do

No

Have you enough money? — No → Ask Mum or Dad for some money

Yes

Do they give you the money ? — No

Yes

Go to the pictures

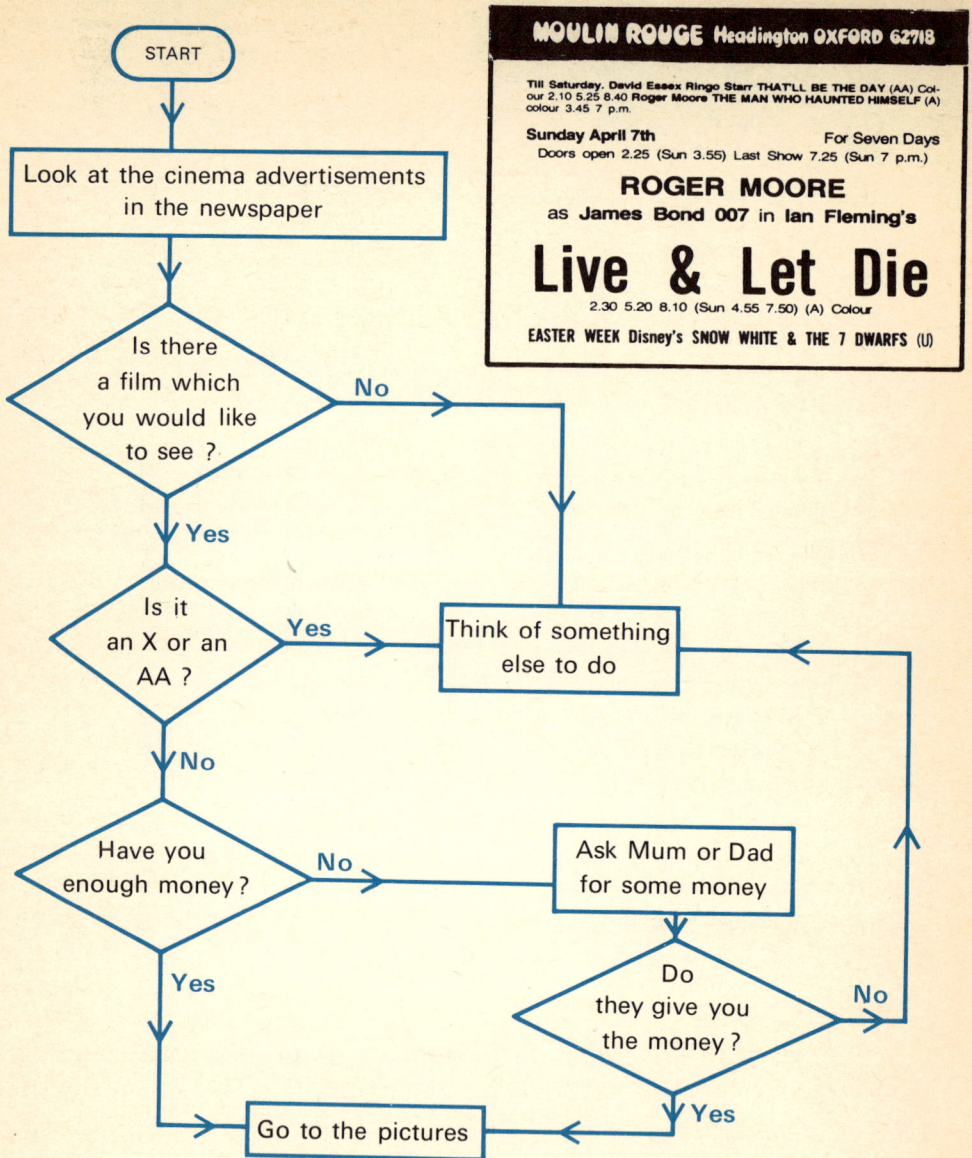

Here is a flow chart which will help you to decide whether you can go to the pictures.

A   Look at the flow chart above.
What shape are: 1. the instruction boxes? 2. the question boxes?

B   Would you change the flow chart in any way?

Preparing clay for pottery

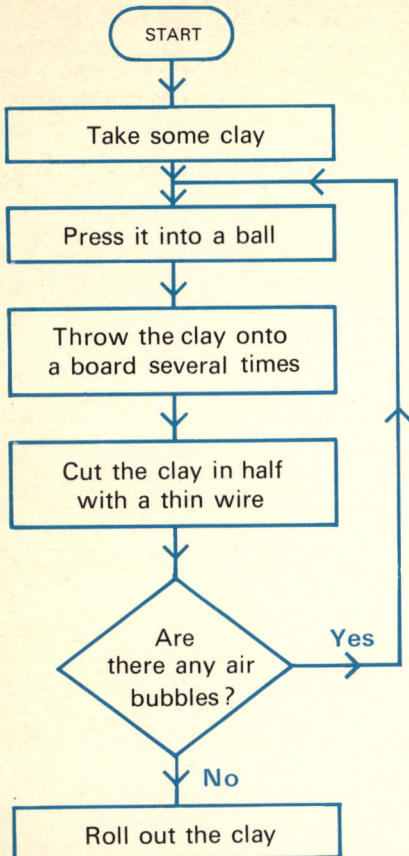

```
          START

    Take some clay

   Press it into a ball

  Throw the clay onto
  a board several times

  Cut the clay in half
  with a thin wire

      Are                Yes
  there any air
   bubbles?

        No

   Roll out the clay
```

Using a signal-controlled road crossing

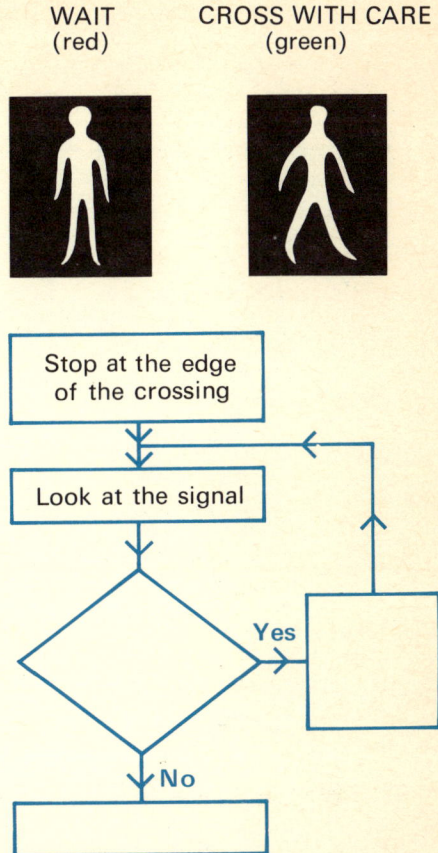

WAIT   CROSS WITH CARE
(red)   (green)

```
  Stop at the edge
  of the crossing

  Look at the signal

                        Yes

        No
```

Sometimes in flow charts a route from a question leads back to an earlier point in the chart. An example is shown on the left above.

A   Copy and complete the flow chart on the right (W/s 1). Remember that there is a push button at the crossing.

B   Draw a flow chart (which includes a question box) to show how to:
1. buy a birthday present for your mother
2. borrow a particular book from your public library
3. make a cup of tea
4. find out why an electric lamp is not working.

START

Draw any triangle

Measure each angle and the length of each edge

Has the triangle three edges the same length? — **Yes** → Write down **equilateral**

↓ **No**

Has the triangle two edges the same length? — **Yes** → Write down **isosceles**

↓ **No**

Write down **scalene** → Has the triangle a right angle? — **No**

↓ **Yes**

Write down .... → STOP

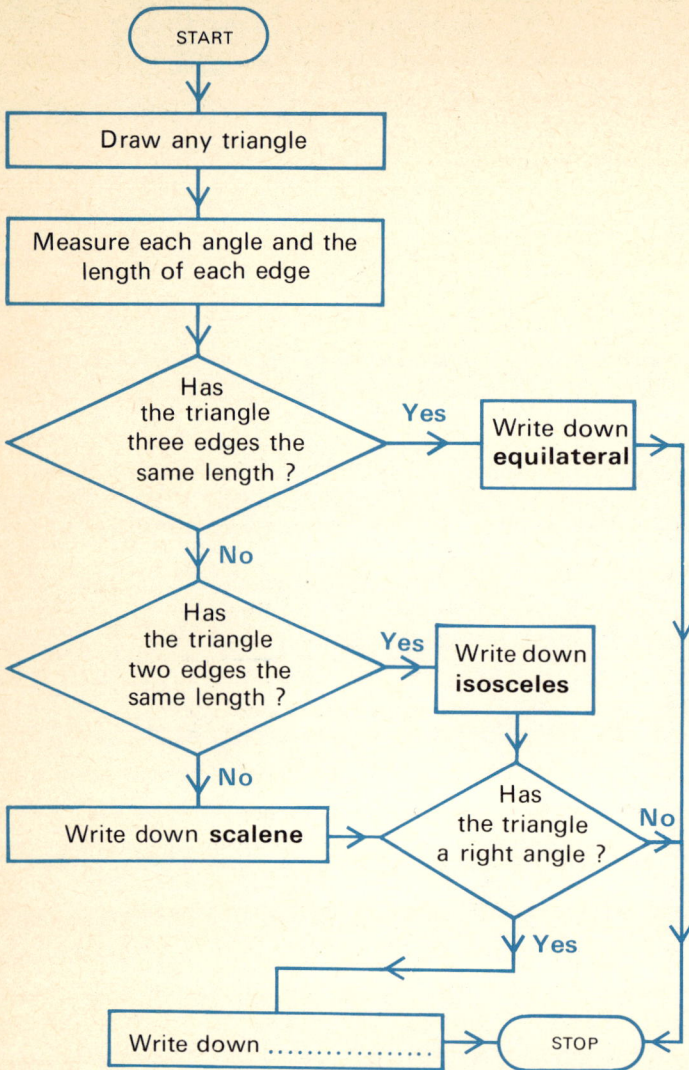

A   Use the flow chart above with the four different triangles shown.
Complete the last instruction box.
Do you agree with the descriptions of your triangles?

B   Use the flow chart with the
triangle shown on the right.
What sort of triangle is it?
Is the flow chart correct?

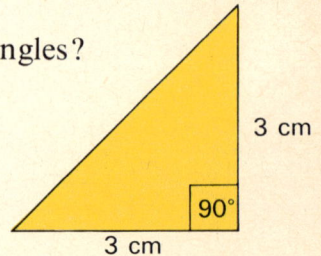

3 cm

90°

3 cm

C   Make up a flow chart to give the descriptions of different numbers.

A Write down the next number in the set:
1. 1, 3, 5, 7, 9, 11, ...
2. 2, 4, 6, 8, 10, 12, ...
3. 3, 6, 9, 12, 15, 18, ...
4. 1, 4, 7, 10, 13, 16, ...
5. 2, 5, 8, 11, 14, 17, ...
6. 1, 2, 4, 8, 16, 32, ...

B For each example in A write down:
1. the 8th number in the set
2. the 10th number in the set
3. the 16th number in the set

C Carry out the instructions in the flow chart:

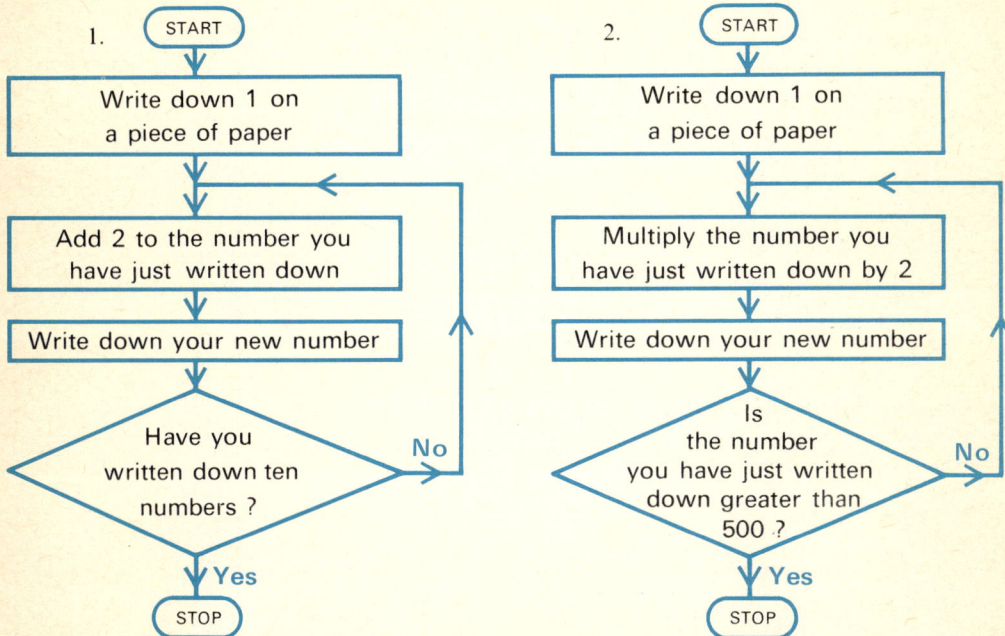

1.
START

Write down 1 on a piece of paper

Add 2 to the number you have just written down

Write down your new number

Have you written down ten numbers? — No

Yes

STOP

2.
START

Write down 1 on a piece of paper

Multiply the number you have just written down by 2

Write down your new number

Is the number you have just written down greater than 500? — No

Yes

STOP

D What was the last number you wrote down in C1?
How many numbers did you write down in C2?
What do you notice about the set of numbers in each case?

E Draw a flow chart for A2, A3, A4, and A5 to give the first 10 numbers in the set.

F Draw a flow chart to give the first 20 numbers in the set:
1, 3, 6, 10, 15, 21, ...

G For each set in A write down the nth number.

Sorting quadrilaterals

You need a pin board, with 9 pins on it arranged in squares as shown below.

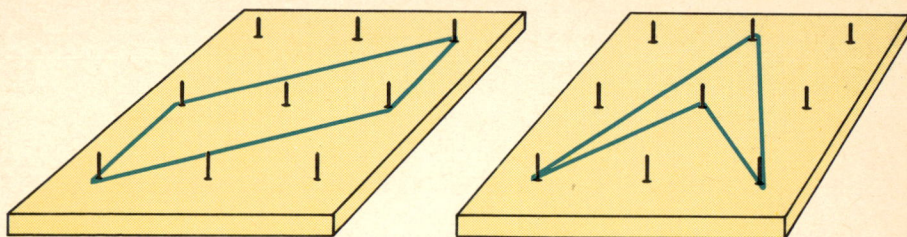

A Make as many different shapes as you can which have four edges. (Remember they must be different in both size and shape and not just be in different positions.)

A shape which has four straight edges is called a **quadrilateral**. Two quadrilaterals are shown at the top of this page.

B Record each new quadrilateral you make on a piece of squared paper (W/s 2, 3).

C Sort your quadrilaterals into various subsets. Record your results in the way you like best.

D Look at the shape on the left at the top of the page. What can you say about this quadrilateral? What do you notice about: 1. its edges?   2. its angles?

A quadrilateral which has each pair of opposite edges parallel is called a **parallelogram**.

E How many different parallelograms can you make on your pin board? What can you say about the lengths of the opposite edges of each parallelogram? What can you say about the sizes of the opposite angles of each parallelogram?

A   Look at the two shapes shown on the right.
      What can you say about the sizes of the angles
      of each shape?
      What can you say about the lengths of the edges
      in the second shape?

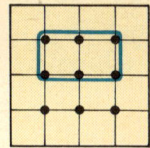

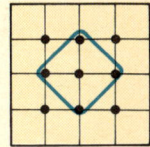

      A quadrilateral in which all the angles are
      right angles is called a **rectangle**.
      A rectangle which has all four edges
      the same length is called a **square**.

B   Draw a large copy of the Venn diagram   Quadrilaterals on a 9-pin board
      on the right (W/s 4).   Rectangles
      Draw, in the correct regions, the   Squares
      shapes you made on your pin board.

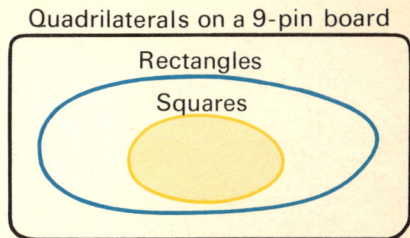

      Squares form a subset of rectangles.
      Since all rectangles have their
      opposite edges parallel they form a subset of the parallelograms.

C   Draw a Venn diagram showing squares, rectangles, and
      parallelograms.

D   Look at the shape shown on the right.
      Two of its edges are parallel.
      How many other quadrilaterals did
      you make with a pair of parallel edges?

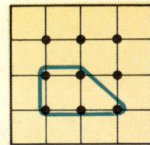

      A quadrilateral which has a pair of parallel edges is called
      a **trapezium**.
      A parallelogram is a special trapezium. It has both pairs of
      opposite edges parallel.

E   Draw a Venn diagram showing squares, rectangles,
      parallelograms, and trapeziums.

F   Look at the last shape on the right.
      What can you say about the lengths of the edges?
      How many lines of symmetry has this shape?
      Can you draw another shape which has these properties?

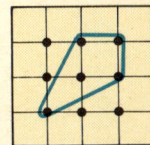

      A quadrilateral which has two pairs of adjacent edges the
      same length is called a **kite**.

Parallelograms

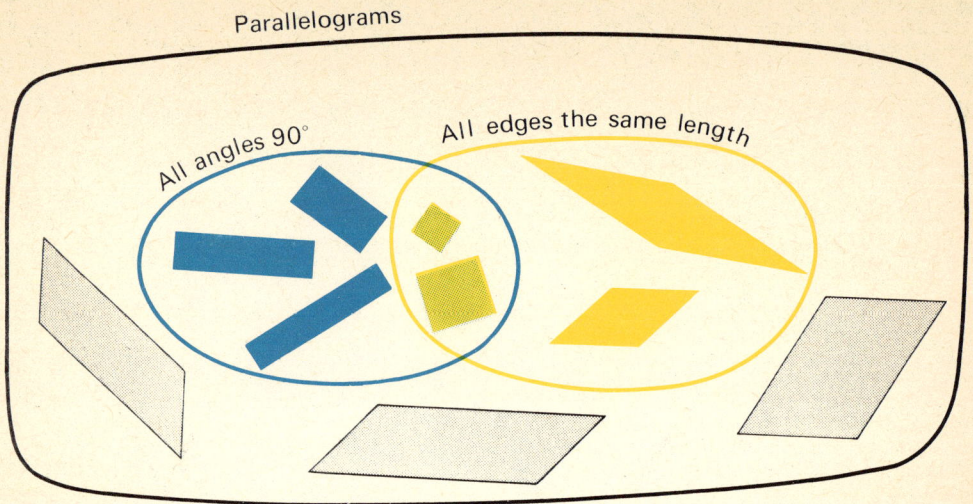

A   Look at the shapes above.
     Can you describe in a single word:
     1. the subset of parallelograms with all angles 90°?
     2. the subset of parallelograms with all edges the same length?

     A quadrilateral which has all its edges the same length is
     called a **rhombus**.

B   How many lines of symmetry has a rhombus?
     Draw a rhombus with edge 4 cm. Mark in the lines of symmetry.
     What can you say about the angles of the rhombus you have drawn?

C   Look at the set of parallelograms at the top of the page.
     What name is given to the parallelograms which belong both to the
     set of rectangles and to the set of rhombuses?

D   Draw a Venn diagram to show a set of quadrilaterals with a
     subset of parallelograms and also a subset of kites.
     Describe the shapes you have placed in the intersection of
     these two subsets.

E   In which region of your Venn diagram in D would you place:
     1. a square?   2. a rectangle?   3. a trapezium?   4. a rhombus?

F   Look for different quadrilaterals inside and outside your classroom.
     Draw each of them and say where you found it.

Does the shape
have at least
one right angle?

Does the shape have
its adjacent edges equal?

if **Yes** go **Left**
if **No** go **Right**

Does the shape have its
opposite edges equal?

trapezium

rectangle

rhombus

square

parallelogram

kite

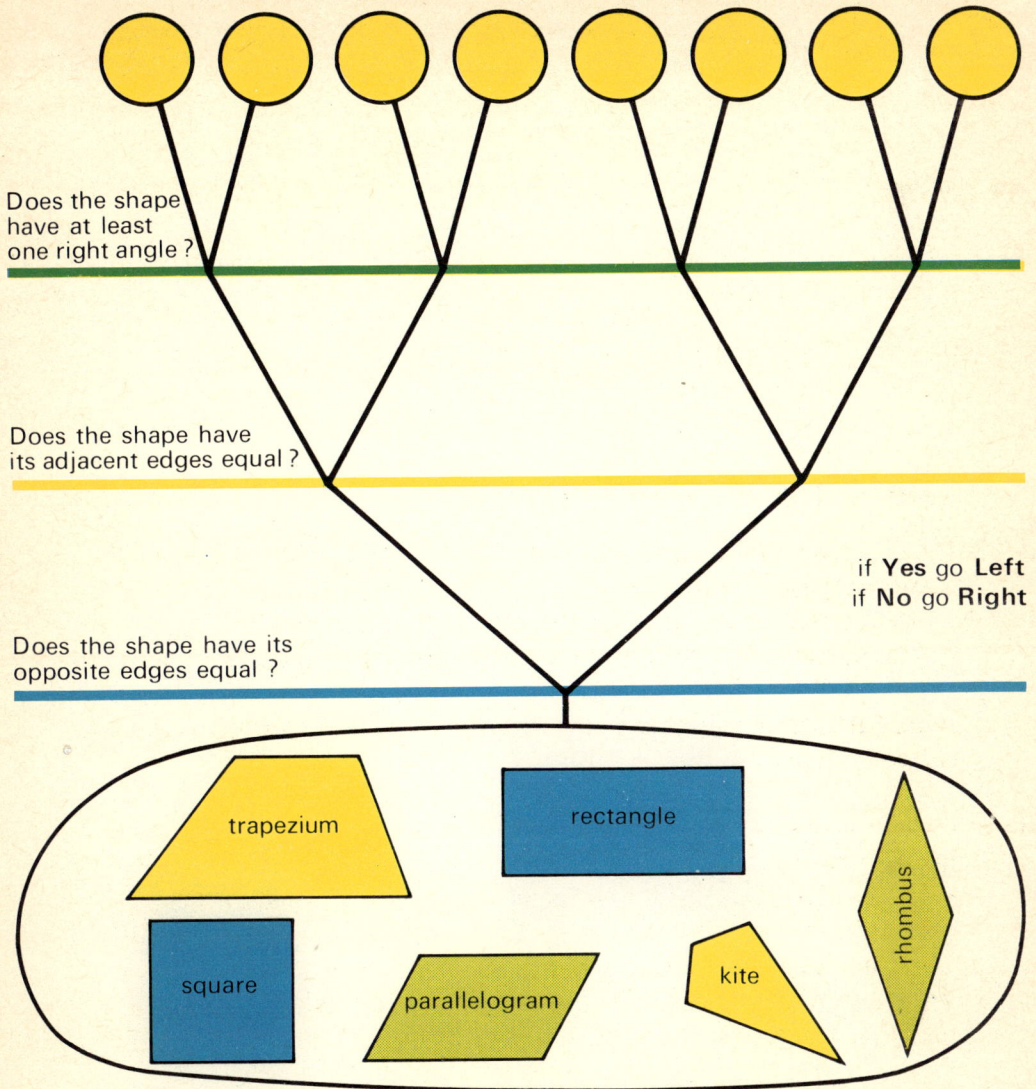

A   Use the sorting tree above to find which shapes finish at the
     end of each branch (W/s 5).

B   You probably found in A that there were no shapes at the end of
     two branches of the tree. Try to draw shapes to fill these two gaps.

C   Draw a rhombus with a right angle and edges 4 cm. What can you
     say about the other three angles? On which branch would it fit?

A Make a drawing of the quadrilateral sketched on the right. Compare your drawing with those of your friends. Are all the drawings the same?

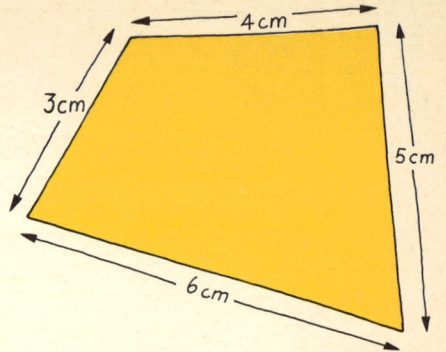

B What extra information would need to be given in A to make sure that all the drawings are the same?

C Make an accurate drawing of each of the parallelograms sketched on the right. Measure the angles in each case. What do you notice? What is this special sort of parallelogram called?

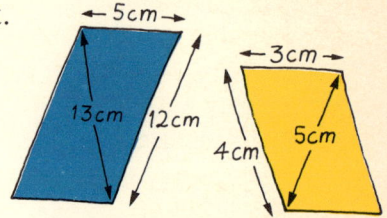

D Make an accurate drawing of the rhombus sketched on the right. Draw the line PQ. Measure PQ. What do you notice? How many lines of symmetry has the rhombus?

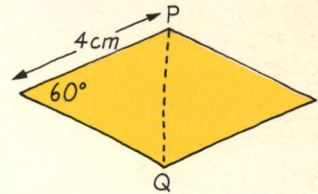

E Make an accurate drawing of the kite sketched on the right. Join PR and QS. Measure the angles at the intersection of PR and QS. What do you notice? Mark in any lines of symmetry.

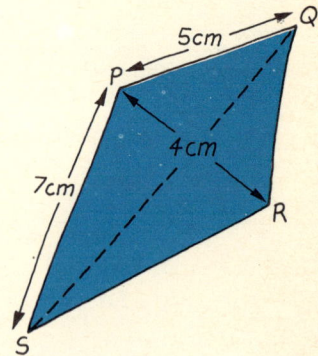

F Draw a trapezium whose parallel edges are 4 cm and 9 cm, and whose other edges are each 5 cm. Measure each of its angles. Explain why two of the angles are the same as those in an equilateral triangle.

G On your drawing of the quadrilateral in A, mark the mid-points of each edge. Join these to form another quadrilateral. Try this with other quadrilaterals. What do you notice?

A   Make a cardboard quadrilateral with its edges all different in length. Make a tessellation using your quadrilateral (W/s 7).

B   Try to describe in your own words how you moved your quadrilateral to make your tessellation.

C   Look at the tessellation at the top of the page.
Make a copy of the yellow quadrilateral (W/s 7).
Describe how you would have to move the yellow quadrilateral in order to fit it onto:
1. the blue quadrilateral   2. the green quadrilateral.

D   Try to fit each angle of the yellow quadrilateral onto one of the four angles at P.
Make a note of which angle fits each angle at P.
What is the sum of the angles at P?
What can you say about the sum of the four angles of the yellow quadrilateral?

The sum of the four angles of a quadrilateral is 360°.

E   Why is the sum of the four angles of a rectangle 360°?

F   What can you say about the sum of two adjacent angles in a parallelogram?

Averages

A   Susan, Jane, Sandra, Kate, Emma,
and Liz work as waitresses
during the holidays.
They decide to share their
tips equally among themselves.
One day the amounts they
received were:

Susan	80p	Jane	90p
Sandra	50p	Kate	100p
Emma	40p	Liz	60p

1. What was the total
   amount they received?
2. How much did they each
   get when they shared
   the money equally
   among themselves?

B   What does this graph
show?

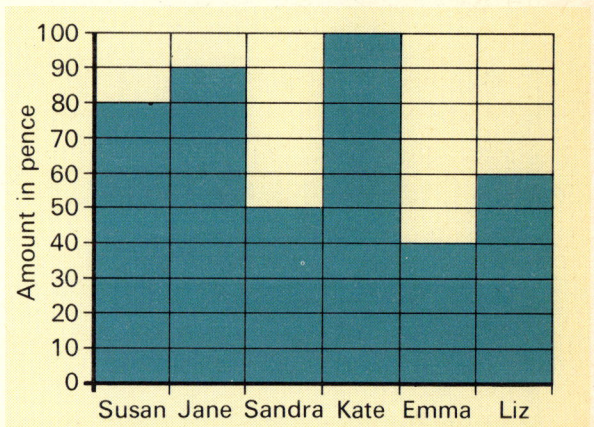

C   What does this graph
show?

D   Are the same total
amounts of money
shown on each graph?

E   Can you see a quick way
of checking that the
amounts of money shown
on each graph are the
same?

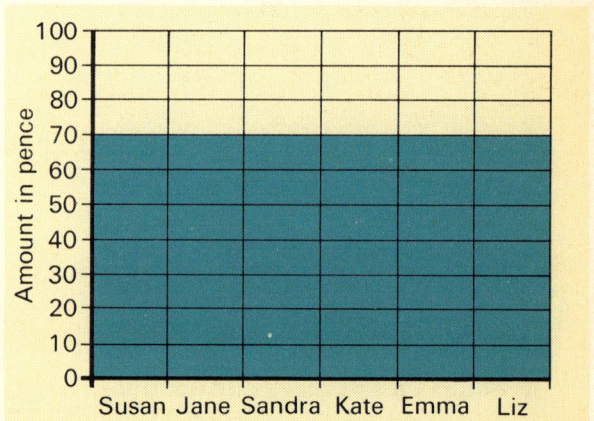

A   The graph shows the number of goals scored by a football team in its first ten matches. Find the total number of goals scored.
Could the same total for the ten matches be obtained in other ways?
Show one way by a graph (W/s 6).
Draw a graph (W/s 6) to show how the same total is obtained if the same number of goals are scored in each match.

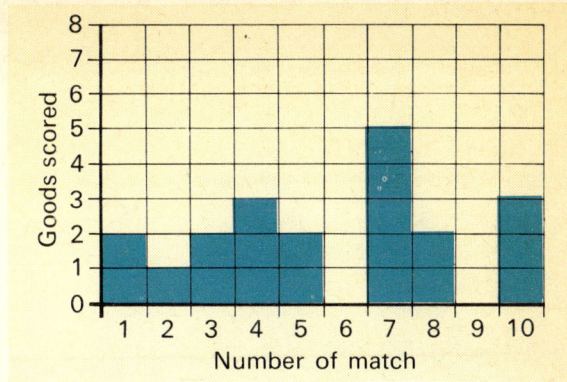

B   Mary's father has his office at home and there are always a lot of letters each day. Mary fetches them from the front door and one week she kept a record of the numbers for each day. Here it is:

Monday	Tuesday	Wednesday	Thursday	Friday	Saturday
27	36	38	26	33	32

Draw a graph (W/s 8) to show these numbers.
Find the total number of letters for the week.
If, for this total, the number of letters were the same on each of the six days, decide on a way of showing this number on your graph.

C   Peter threw four dice of different colour and scored a total of 20 as shown.
In what other ways could he get the same total?
In which of these would the numbers on the dice all be the same?

D   The quadrilateral shown here is made of wire.
It is made into a square.
What is the length of each edge of the square?

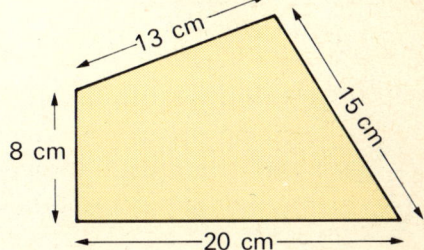

13 cm

15 cm

8 cm

20 cm

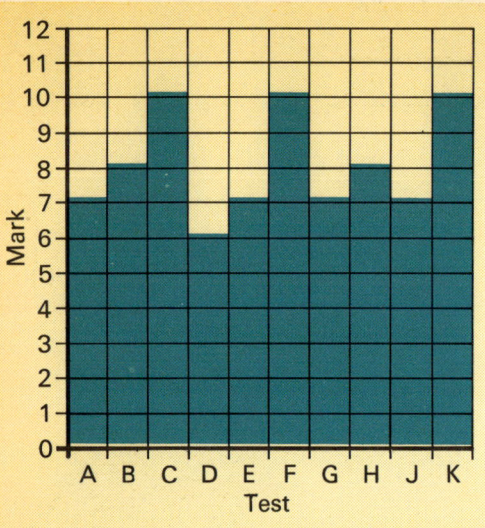

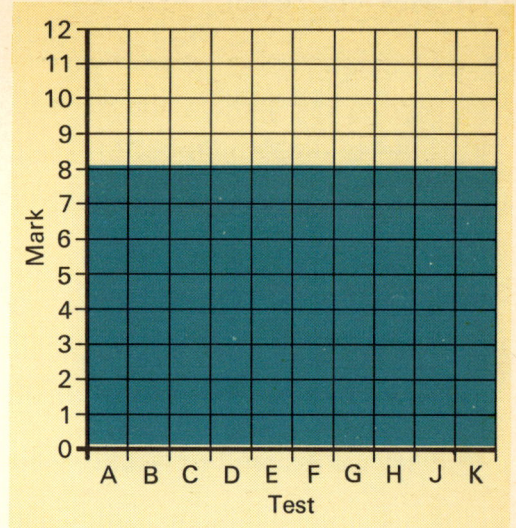

A   Look at the graph on the left above.
     It shows the marks obtained by a boy in some tests.
     Which mark did he have most often (i.e. the mode)?

B   How many tests were there altogether?
     What was the total of all the marks?
     If, for this total, the boy had had the same mark in each test,
     what would it have been?
     What does the second graph show?

     If each member of the set is given the same value (keeping the total
     unchanged) then this value is called the **mean** of the set.

C   Look back to page 16.
     What is the mean of the money received by the six girls?

D   Look back to page 17. What is the mean of:
     1. the number of goals scored by the football team?
     2. the number of letters received each day by Mary's family?
     3. the numbers on the four dice?
     4. the lengths of the edges of the quadrilateral?

     The mode and the mean each help to describe a set.
     They are each used to give an *average* value of the members of
     a set. Sometimes the mode is used. Sometimes the mean is
     more helpful.

A   Measure, in cm, the lengths of the five lines.

1. _____

2. _____

3. _____

4. _____

5. _____

Find the mean length of the lines.

B   Kate is a good swimmer and
goes to the baths for special
coaching.
On her last visit to the baths
she swam four separate lengths
and was timed for each.
They were:
27·9s, 27·5s, 26·9s, 27·3s.
Find Kate's mean time
for a length.

C

Hours of sunshine						
Sat.	Sun.	Mon.	Tues.	Wed.	Thurs.	Fri.
8·6	9·5	7·5	2·3	6·9	8·8	9·6

The table above shows the hours of sunshine for a week at
Torquay in July.
Find the mean hours of sunshine each day.
The mean hours of sunshine each day in July over the
last 30 years in Torquay is 7·2.
Was the week shown above a good week for sunshine?

D   Find the mean amount of money which you and your friends
have with you at present.

E   Find a large stone. With some friends, each estimate the mass
of the stone in grams. Find the mean of your estimates.
Now weigh the stone.
Do you think that you and your friends are good at estimating
mass?

A   Philip was telling his friend
     about a car journey from
     Leicester to Bournemouth.
     'We went over 100 km per hour
     on some roads' he said. 'At
     other times, however, we just
     crawled along. But we managed
     to cover the 240 km in 4 hours.
     We travelled at an average
     speed of 60 km per hour.'
     Explain what is meant by
     average speed.

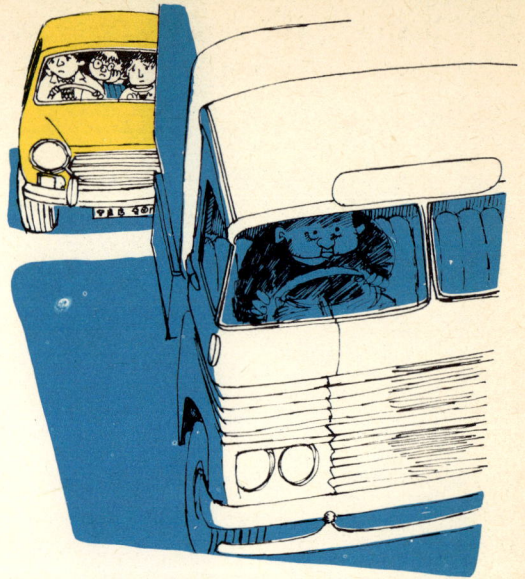

B   What is the average (mean) speed of a car if it travels:
     1. 200 km in 5 hours?   2. 450 km in 9 hours?
     3. 320 km in 4 hours?   4. 315 km in 7 hours?

Two ways of showing the working for finding the average (mean)
speed of a car which travels 520 km in 7 hours are given below.

$$
\begin{array}{r}
520 \\
-490 \\
\hline
30 \\
-28 \\
\hline
20 \\
-14 \\
\hline
60 \\
-56 \\
\hline
40 \\
-35 \\
\hline
5
\end{array}
\qquad
\begin{array}{l}
(7 \times 70) \\
\\
(7 \times 4) \\
\\
(7 \times 0.2) \\
\\
(7 \times 0.08) \\
\\
(7 \times 0.005)
\end{array}
$$

tenths
hundredths
thousandths

$$
\begin{array}{r}
74.285 \\
7\overline{)520} \\
490 \\
\hline
30 \\
28 \\
\hline
20 \\
14 \\
\hline
60 \\
56 \\
\hline
40 \\
35 \\
\hline
5
\end{array}
$$

The division does not work out exactly. $520 \div 7 = 74.285\ldots$

To the nearest whole number $520 \div 7 = 74$
To the first place of decimals $520 \div 7 = 74.3$
To the second place of decimals $520 \div 7 = 74.29$

C   Find the answer to the first place of decimals:
     1. $470 \div 7$ 2. $520 \div 6$ 3. $475 \div 8$ 4. $590 \div 9$

Two ways of showing the working for finding the average (mean) speed of a car which travels 1070 km in 14 hours are given below.

```
                76·42                                        76·42
           14)1070                                       14)1070
               980        (14 × 70)                          980
                90                                            90
                84        (14 ×  6)                           84
tenths          60                                           60
                56        (14 × 0·4)                          56
hundredths      40                                           40
                28        (14 × 0·02)                         28
                12                                           12
```

The average (mean) speed is 76·4 km/h, to the first place of decimals.

A   Find, to the first place of decimals, the average (mean) speed of a car which travels:
1. 1066 km in 13 hours   2. 1100 km in 15 hours
3. 2000 km in 23 hours   4.   967 km in 9 hours

Name	Runs	Completed innings
D. B. Close	1096	32
J. H. Edrich	1039	34
A. W. Greig	721	27
A. P. E. Knott	719	28
B. W. Luckhurst	1141	33
R. M. Prideaux	955	36

B   Find, to 2 places of decimals, the average (mean) score of each of the cricketers in the table above.

C   Ask a friend to measure your pulse rate. This means counting the number of times your heart beats in one minute.

D   Work with about twelve friends. Find each other's pulse rates. Record the results. Find the mode and the mean.
Now, in turn, find each other's pulse rates after jumping on and off a chair 25 times. Again find the mode and the mean.
Compare the two sets of results. Write down all you find out.

Early bicycles did not have
tyres like those we have now.
They had an iron band round
each wheel. The bicycles were
heavy.
Later, solid rubber was used
instead of the iron. The bicycles
then went a little more smoothly.
At the end of the last century
Mr. Dunlop started to use
pneumatic tyres. The use of these
tyres made cycling much easier
and smoother and they quickly
became very popular.

A Look at the paragraph above.
Look at the long words.
Look at the short words.
Count how many words
there are with 3 letters.
Count how many there are
with 4 letters, 5 letters, etc.
Show your results on a
graph (W/s 9), like that
started on the right.
(There are 16 words
with 3 letters.)

B Look at your graph.
What is the mode?
What is the total number
of words? How did you
find the total?
What is the total number
of letters? How did you
find the total?

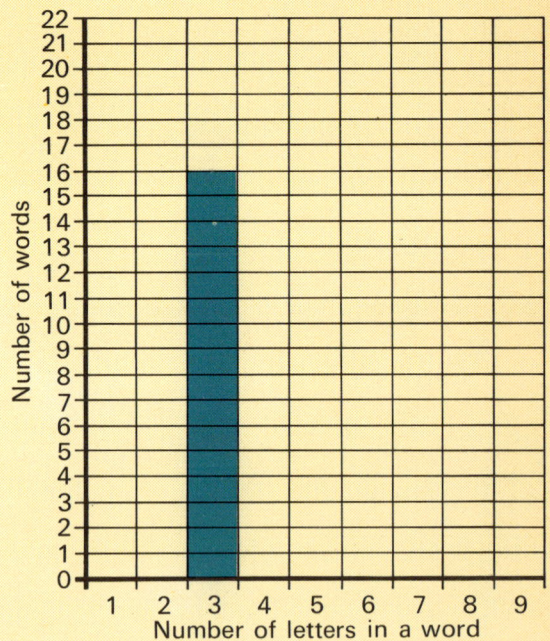

C Find, to the first place of decimals, the mean number of letters
in each word.

D Repeat A, B, and C for a paragraph from a book or newspaper
(W/s 10).

A   Obtain about twenty boxes
    of the same brand of matches.
    What is the stated average
    contents?
    Investigate whether this
    number refers to the mode
    or to the mean (or both).

B   Obtain information about the shoe sizes of the children in
    your class. Show your results on a graph.
    Find the mode and the mean.
    Which do you think is most useful to use as the average?
    Is the average the same for boys as for girls?

C   Find the mean attendance at the matches
    shown on the right.
    Give your answer:
    1.   to the nearest ten
    2.   to the nearest hundred
    3.   to the nearest thousand
    Which do you think is the most sensible
    way of giving your answer?

Division One

ARSENAL......1	0.....SOUTHMPTN
Ball	19,210
COVENTRY... 0	1.....BIRMINGHM
	Hatton 27,825
DERBY.........1	1.............STOKE
Bourne	Ritchie 28,176
IPSWICH.....1	1........NORWICH
Hamilton	Boyer 25,004
LEEDS..........1	1.....NEWCASTLE
Clarke	Barrowclough
	46,611
LEICESTER...2	1.........EVERTON
Workington	Latchford
Earle	22,286
LIVERPOOL...1	0.........BURNLEY
Toshack	42,562
Q.P.R..........3	1.....TOTTENHAM
Givens,Bowles	Chivers
Francis	25,775
SHEFF.UTD...0	1.........MAN.UTD
	Macari 29,203
WEST HAM....3	0..........CHELSEA
Bonds 3	34,043

D   In ice skating competitions there are several judges who each
    award a mark. They then use a special method for finding the
    mean of these marks.
    Find out all you can about the method used.
    Do you think it is a good method? Why?

E   Find the mean of:
    1. 101, 102, 103, 104, 105   2. all the whole numbers from 1 to 99.

F   Write down the mean of:
    1. x, y, and z 2. p, q, r, and s
    3. a, a, b, b, b 4. m, m, m, n
    5. x, x, y, y, z, z 6. s, t, s, t, p, q, q, q

Intersection

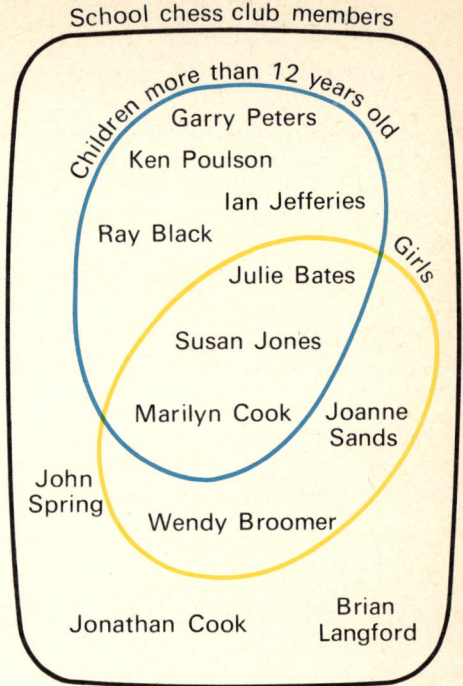

School chess club members

Children more than 12 years old

Garry Peters
Ken Poulson
Ian Jefferies
Ray Black
Julie Bates
Susan Jones
Marilyn Cook   Joanne Sands
John Spring
Wendy Broomer
Girls
Jonathan Cook   Brian Langford

A   The Venn diagram above shows the members of a school chess club.
Using this diagram write down the names of the children who are:
1. girls twelve years old or younger
2. boys more than twelve years old
3. boys twelve years old or younger
4. girls more than twelve years old

The set of girls more than twelve years old form the intersection
of the set of girls *and* the set of children more than twelve years old.

To show which members belong to a set, curly brackets {   }
are often used, as in B below.

B   Write down the members of the set formed by the intersection of:
1. {2, 4, 6, 8, 10, 12, 14, 16, 18, 20} *and* {3, 6, 9, 12, 15, 18}
2. {1, 2, 3, 4, 6, 12} *and* {1, 2, 4, 8, 16}
3. {(1, 4), (2, 5), (3, 6), (4, 7), (5, 8)} *and* {(1, 2), (2, 4), (3, 6), (4, 8)}

C   Can you describe each of the sets, mentioned in B, in another way?
Describe the intersection set in B1 and in B2.

The symbol used for intersection is ∩.
Using this symbol, exercise B2 on the opposite page can be
written as: {1, 2, 3, 4, 6, 12} ∩ {1, 2, 4, 8, 16} = {1, 2, 4}.

A   Write down the members of the set formed by:
     1. {1, 4, 7, 10, 13, 16, 19, 22, 25} ∩ {1, 4, 9, 16, 25}
     2. {6, 12, 18, 24, 30, 36} ∩ {3, 6, 9, 12, 15, 18, 21, 24, 27, 30, 33, 36}

B   What do you notice about your answer to A2? Can you say why?

C   Draw a Venn diagram for each example in A.
     Colour the region which shows the intersection.

Shapes

D   Look at the shapes above.
     Make a copy of: 1. {shapes which have some straight edges}
                       2. {shapes which have some curved edges}

E   Draw the set of shapes formed by the intersection of the two
     sets you found in D.

F   Show the information from D on a Venn diagram.
     Colour the region which shows the intersection.

G   Describe the shapes in the non-coloured regions of your Venn
     diagram in F.

H   Find the set of playing cards formed by
     {red cards} ∩ {picture cards where the person has only one eye}.

A   From the children in your class write down:
   1. {children who come to school by public transport}
   2. {children who stay to lunch regularly}
   3. {children who ride a bicycle to school}

B   Try to show the information in A on a Venn Diagram.

It is often convenient to describe a set by using a capital
letter. This saves having to write out the full description.
In the first question on this page you could use P to stand
for the set of children who come to school by public transport. In the
same way you might use L for the set of children who stay to
lunch regularly, and B for those who come by bicycle.

C   If the letters P, L, and B stand for the sets in exercise A find:
   1. $P \cap L$ 2. $P \cap B$ 3. $L \cap B$ 4. $P \cap L \cap B$.

D   Describe in words $P \cap L \cap B$.

E   If W stands for {whole numbers from 1 to 40},
        Q stands for {factors of 24},
        R stands for {factors of 27},
        S stands for {factors of 36},
   write down the members of each set.

F   Show the sets W, Q, and R, in exercise E on a Venn diagram.
   Find $Q \cap R$.
   Which is the largest number in the intersection?

Is the number a factor of 36?

Is the number a factor of 27?

If **Yes** go **Left**
If **No** go **Right**

Is the number a factor of 24?

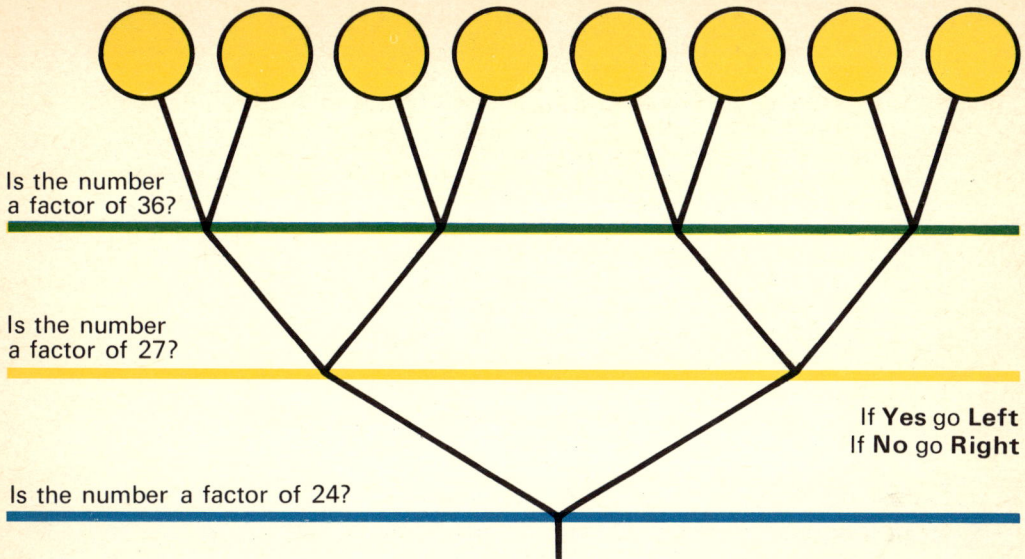

A   Use the sorting tree above to sort {whole numbers from 1 to 40} (W/s 11).

B   Describe in your own words the subset of numbers which finish:
     1. on the extreme left-hand branch
     2. on the extreme right-hand branch

C   Using the information in exercise E on page 26, complete the Venn diagram (W/s 12) shown on the right.

Whole numbers from 1 to 40

Factors of 24 (Q)

Factors of 27 (R)

Factors of 36 (S)

D   What do you notice about the eight subsets you found in exercise A and the numbers in the eight regions of your Venn diagram? Can you say why?

E   On your Venn diagram colour Q ∩ R red, Q ∩ S blue, and R ∩ S yellow.
     Which region represents Q ∩ R ∩ S?
     Which is the largest number in this region?

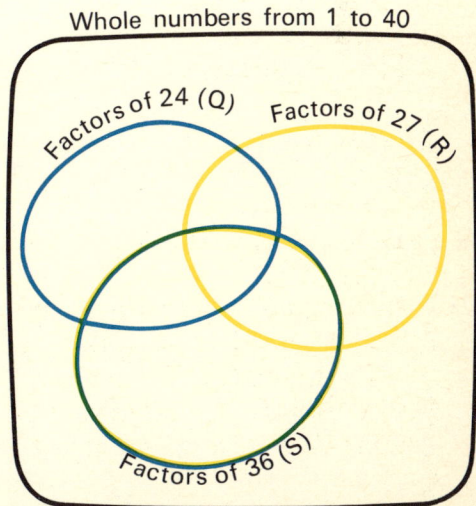

Whole numbers from 1 to 20

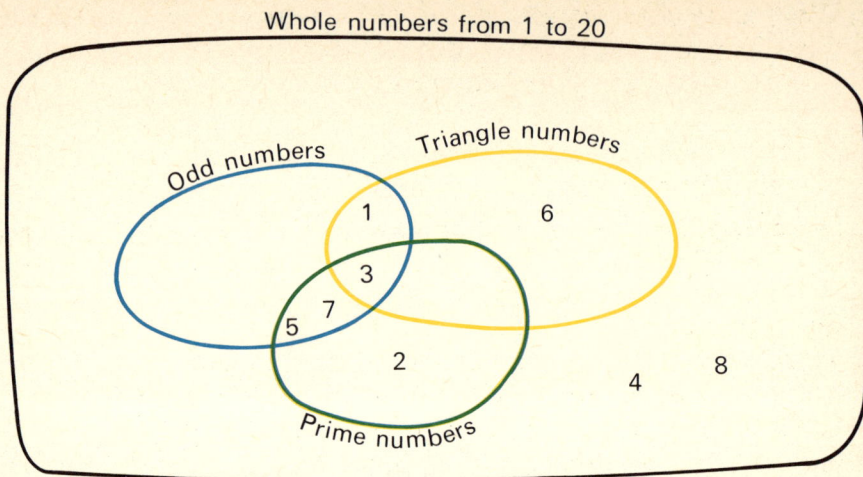

A | Copy and complete the Venn diagram above (W/s 13).

B | Write down the members of the set formed by:
   1.   {triangle numbers} ∩ {odd numbers}
   2.   {triangle numbers} ∩ {prime numbers}
   3.   {odd numbers} ∩ {prime numbers}

Whole numbers from 1 to 40

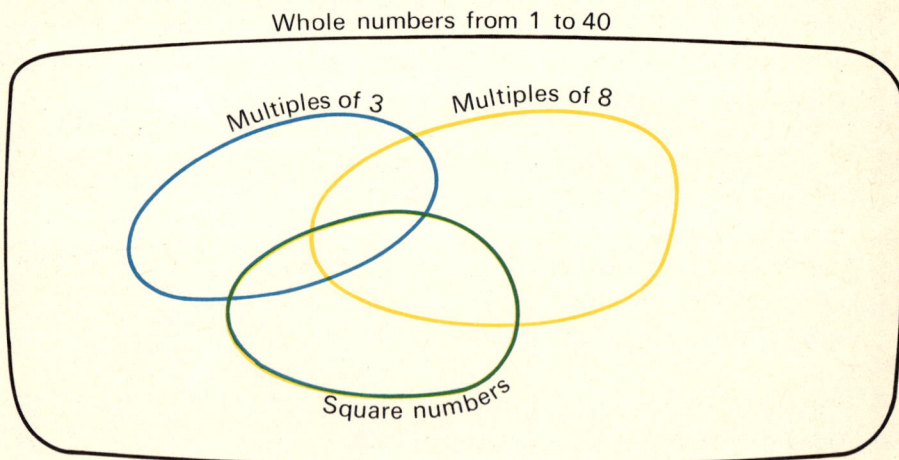

C | Copy and complete the Venn diagram above (W/s 14).

D | Write down the members of the set formed by:
   1.   {multiples of 3} ∩ {multiples of 8}
   2.   {multiples of 3} ∩ {square numbers}
   3.   {multiples of 8} ∩ {square numbers}

A   What is the intersection of the set of points 5 cm from a fixed point P, and the set of points forming a straight line through P?

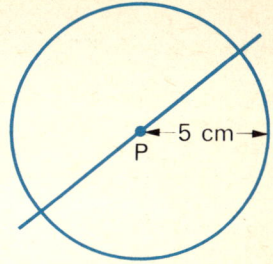

B   What is the intersection of the set of points less than 5 cm from a fixed point P, and the set of points forming a straight line through P?

C   Draw two lines which intersect at right angles.
Mark a set of points 3 cm from one of these lines.
Mark a set of points 4 cm from the other line.
What is the intersection of these two sets?
How would you describe these points?

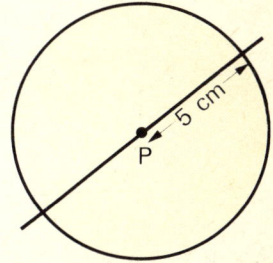

D   Draw a triangle PQR on a piece of plain paper.
Mark a set of points which are the same distance from the points P and Q.
Mark a set of points which are the same distance from the points P and R.
What is the intersection of these two sets?
Mark a set of points which are the same distance from the points Q and R.
What is the intersection of the three sets of points you marked?
What can you say about this point?
Is it possible to draw a circle passing through P, Q, and R?

E   Draw a triangle LMN on a piece of plain paper.
Mark the set of points which are the same distance from the lines LM and LN.
Mark the set of points which are the same distance from the lines ML and MN.
What is the intersection of these two sets?
Mark the set of points which are the same distance from the lines NM and NL.
What do you notice?
Draw a circle which touches LM, MN, and NL.

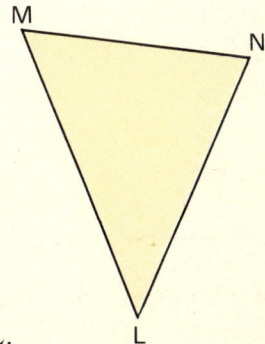

Area and volume

A   Find the area of the rectangle.
What fraction of the rectangle is the
coloured triangle?
Write down the area of the triangle.

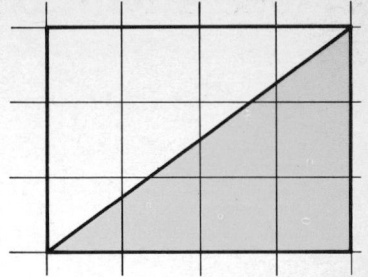

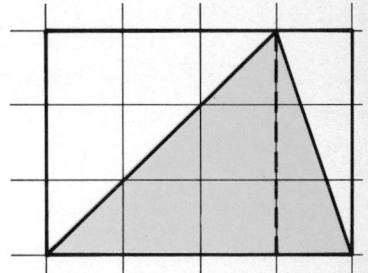

B   Find the area of the coloured triangle.

1.   2.

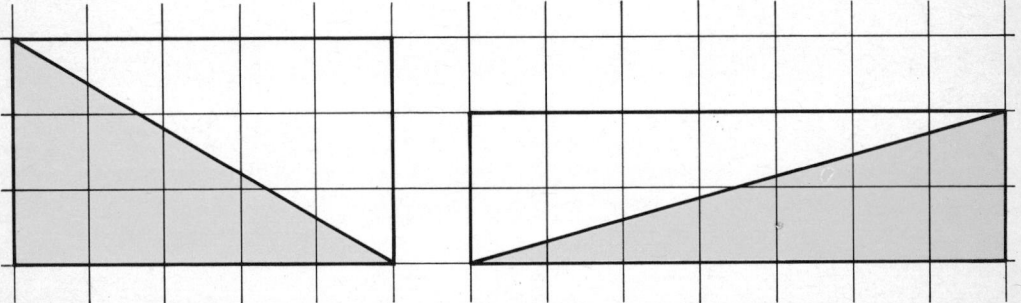

C   What is the area of the rectangle?
What fraction of the rectangle is the
coloured triangle?
Write down the area of the triangle.

D   Find the area of the triangle.

1.   2.

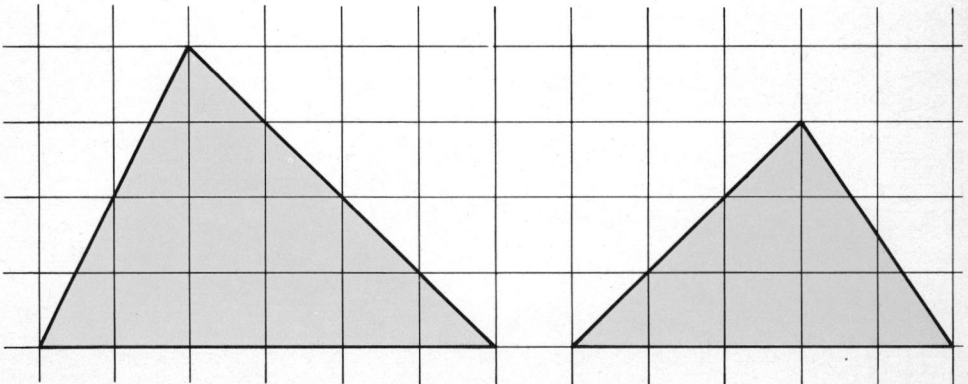

E   Write, in words, how you would find the area of:
1. a rectangle   2. a triangle

A   Find the area of the triangle:

1.

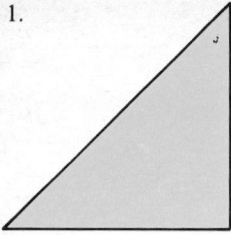

2.

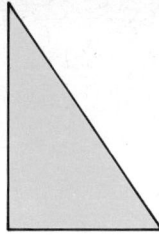

3.

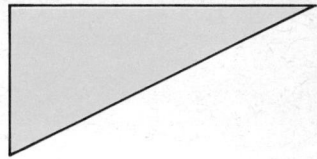

4.

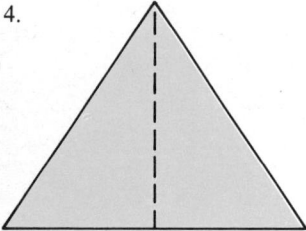

5.

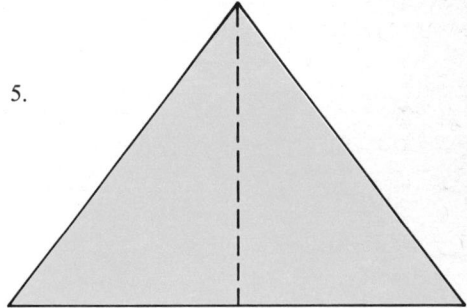

B   Find the area of the triangle:

1.

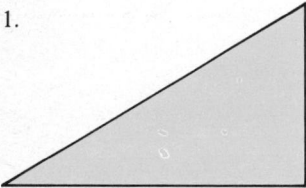

2.

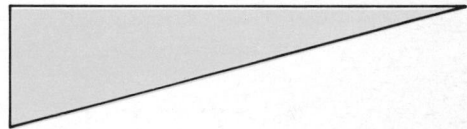

3.

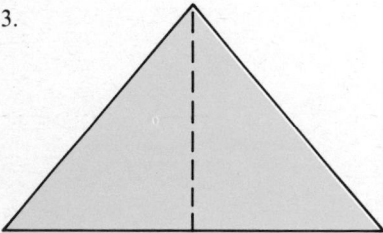

4.

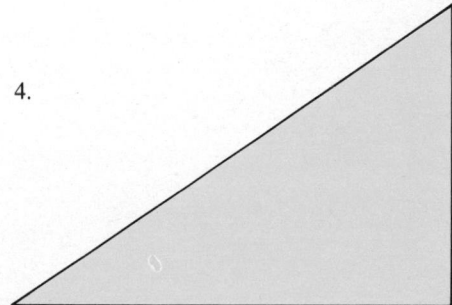

A   Find the area of the coloured
     triangle.
     Use your result to find the
     area of the regular hexagon.

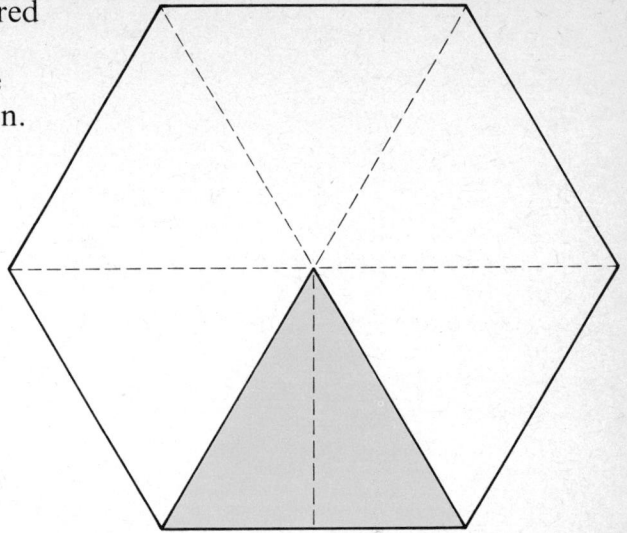

B   Find the area of the regular polygon.

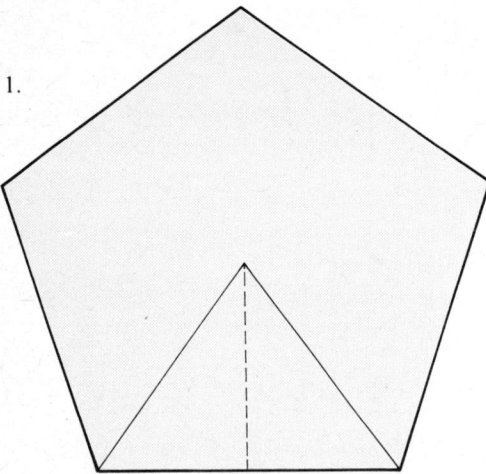

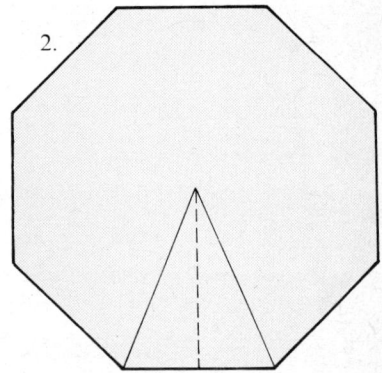

1.

2.

C   Copy and complete the table about regular polygons (W/s 15).

Number of edges	6	5	8	7	12
Length of each edge	8 cm	6 cm	4 cm	4 cm	2 cm
Distance of each edge from the centre	6·9 cm	5·1 cm	4·8 cm	4·2 cm	3·7 cm
Area in cm²					

A Look at the rectangle.
What are its length and width, in cm?
What are its length and width, in mm?
What multiplication would you try to
do to find the area in cm²?
Can you find the area in mm²?
How would you change your answer to cm²?
Find the area in cm².

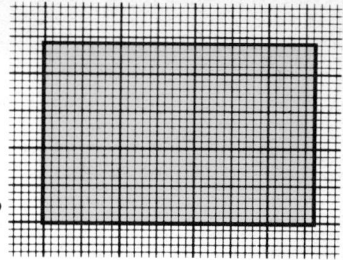

Here are some possible answers to A.

Length = 3·6 cm   Width = 2·4 cm   Area = 3·6 × 2·4 cm²
Length = 36 mm   Width = 24 mm   Area = 36 × 24 mm²

$$\begin{array}{r} 36 \\ \times\,24 \\ \hline 144 \\ 720 \\ \hline 864 \end{array}$$

Area = 864 mm²   100 mm² = 1 cm²
Area = 8·64 cm²

3·6 × 2·4 = 8·64

B Find, in cm², the area of the rectangle:

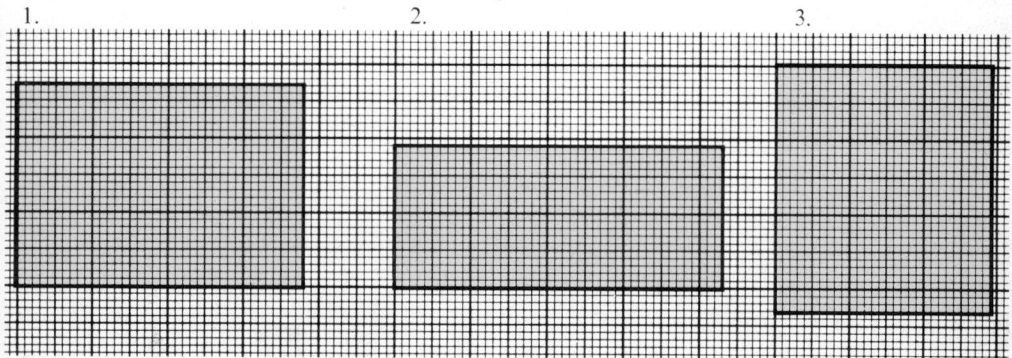

1.   2.   3.

C Copy and complete the table (W/s 15).

Length	4·3 cm	5·6 cm	6·8 cm	9·8 cm	21·7 cm
Width	2·1 cm	3·7 cm	1·7 cm	5·2 cm	16·9 cm
Area in cm²					

D Copy and complete the multiplications. What do you notice?

1. 22 × 14 =   2. 36 × 27 =   3. 62 × 48 =
   2·2 × 14 =   3·6 × 27 =   6·2 × 48 =
   2·2 × 1·4 =   3·6 × 2·7 =   6·2 × 4·8 =

A Using the dimensions shown, find the area of the coloured triangle.

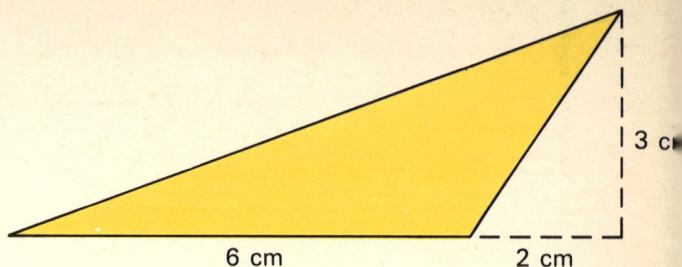

6 cm   2 cm   3 cm

B Find the area of the coloured triangle.

1.

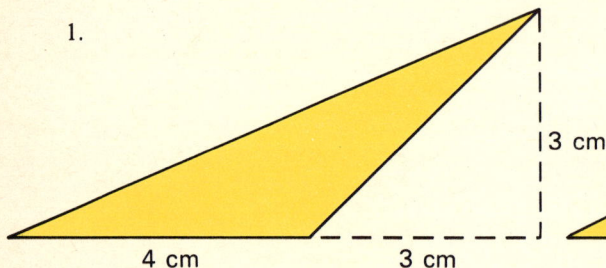

4 cm   3 cm   3 cm

2.

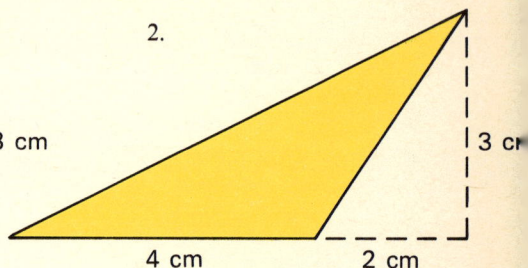

4 cm   2 cm   3 cm

3.

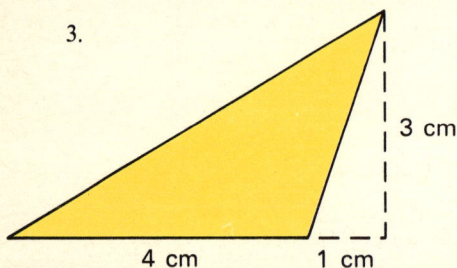

4 cm   1 cm   3 cm

4.

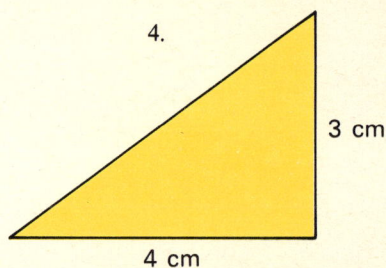

4 cm   3 cm

C What did you notice about the areas of the four triangles in B? Do you think that *all* triangles with a base of 4 cm and a height of 3 cm will have an area of 6 cm²?

D Draw, as above, several triangles with a base of 6 cm and a height of 4 cm. Find the area of each. What do you notice?

E Copy and complete the table for triangles (W/s 15).

Base length Height	8 cm 4 cm	5 cm 7 cm	4·3 cm 4 cm	12 cm 3·7 cm	27 cm 6·7 cm
Area in cm²					

F Write down the area of a triangle with base *b* cm and height *h* cm.

A Find the area of:
(a) the coloured triangle   (b) the parallelogram

1.
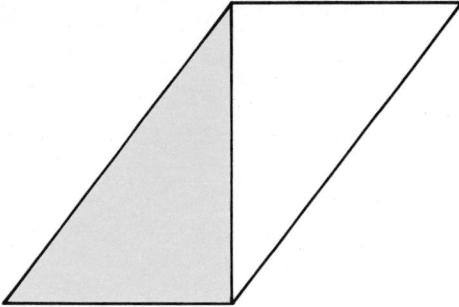

2.
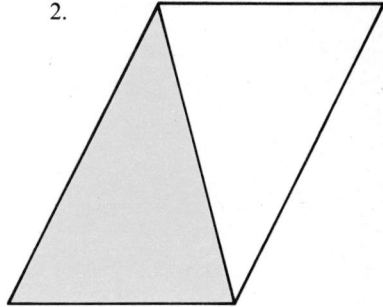

3.
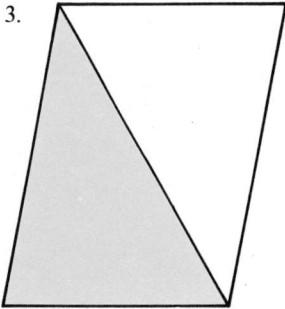

4.
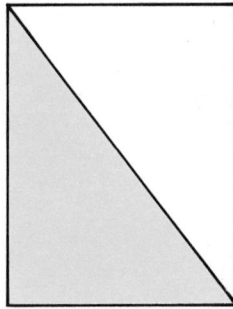

B What do you notice about the areas of the parallelograms in A? Do you think that *all* parallelograms with a base of 3 cm and a height of 4 cm have an area of 12 cm²?

C Draw a set of parallelograms, each of which has a base of 6 cm and a height of 5 cm. Find the area of each.

D Find the area of the parallelogram.

1.
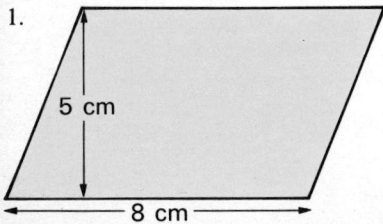

2.
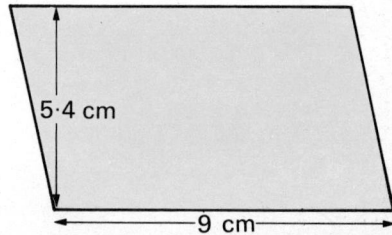

E Write down the area of a parallelogram with base b cm and height h cm.

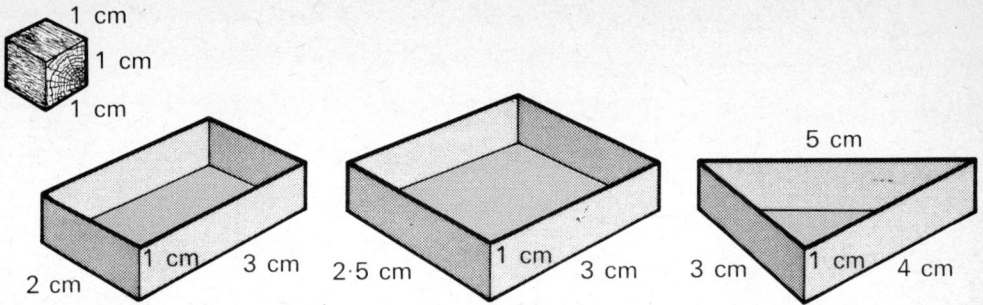

A   Look at the drawings of three trays and a wooden centimetre cube.
     Which of them could you completely fill, using only whole cubes?
     How many cubes would be needed?

B   In A you probably found that you could fill only the first tray.
     If plasticine cubes were used, some could be cut into smaller pieces
     to fill gaps in the other two trays. Use the bases of the three trays,
     shown below, to help you to find how many plasticine cubes would
     be needed to fill each of the three trays.

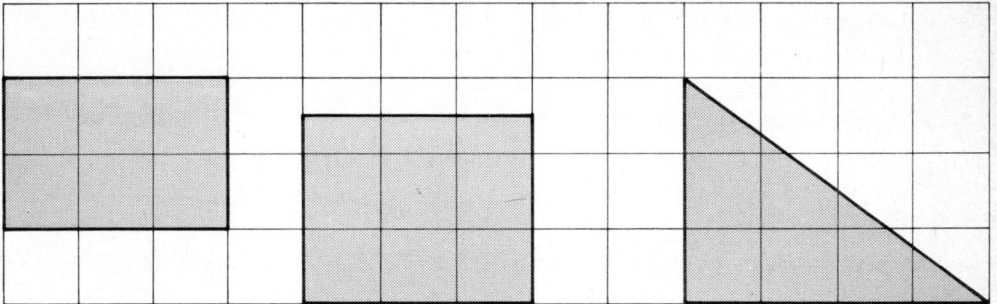

C   If the height of each tray were 2 cm find how many cubes would
     be needed to fill each. How many would be needed if the height
     were 3 cm?

D   Find the volume of   1.
     a shape with the
     dimensions shown.

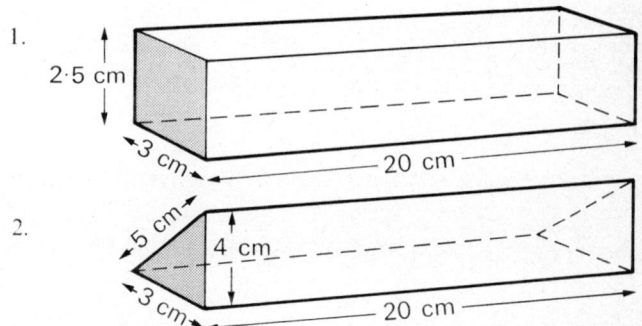

     2.

A Find the volume of a prism with the dimensions shown:

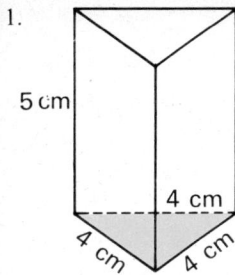

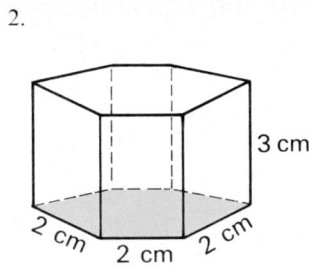

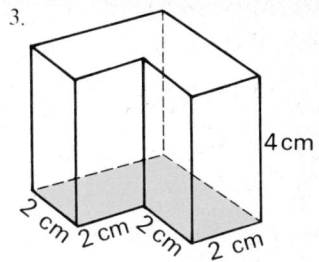

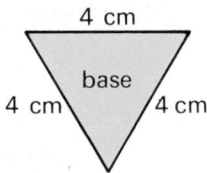

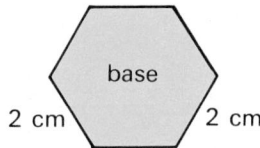

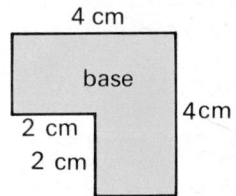

1.

5 cm

4 cm

4 cm   4 cm

4 cm

4 cm   base   4 cm

2.

3 cm

2 cm   2 cm   2 cm

2 cm   base   2 cm

3.

4 cm

2 cm   2 cm   2 cm   2 cm

4 cm

base

2 cm

2 cm   2 cm

4 cm   4 cm

base   4 cm

B Explain how you would find the volume of any prism.

C Collect a set of prisms (cartons, etc.).
Find the volume of each of them.

D An L-shaped steel girder
has the dimensions
shown in the diagram.
Find the volume of the
metal used.
If one cubic centimetre of the steel has a mass of 7·8 g, find
the total mass of the girder.

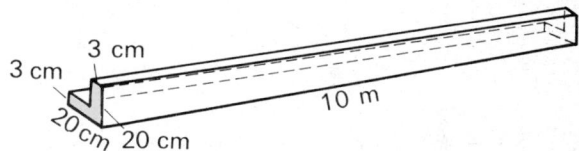

3 cm

3 cm

10 m

20 cm   20 cm

E The volume of a cuboid is 672 cm³. Its length is 7 cm and its
width is 8 cm. Find the height of the cuboid.

F The volume of a prism is 576 cm³. The area of the cross section
is 72 cm³. Find the height of the prism.

G Draw a circle of radius 4 cm.
Draw a regular hexagon with its vertices on the circle.
Find the area of the hexagon. Then find the volume of a prism
with the hexagon as base and 8 cm in height.

Ordered pairs

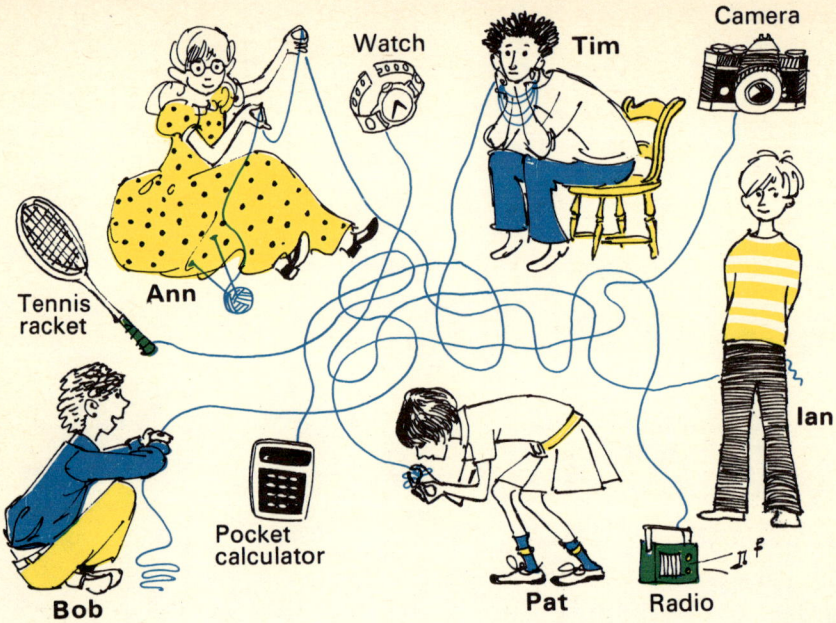

Watch   Tim   Camera

Tennis racket   Ann

Ian

Bob   Pocket calculator   Pat   Radio

A   In the picture above each of the children is going to receive one of the items shown as a present. Which is Ann's present?

B   Show the information in the picture as a set of ordered pairs:
(name of child, present received)

C   The arrow graph on the right shows the result of six football matches. Which team won when Arsenal played Wolves?

D   Show the information in the arrow graph as a set of ordered pairs:
(winning team, losing team)

E   Which team won the most matches? Which team won the least matches?

F   Explain how you can answer E using:
1. the arrow graph
2. the set of ordered pairs in D

won when playing

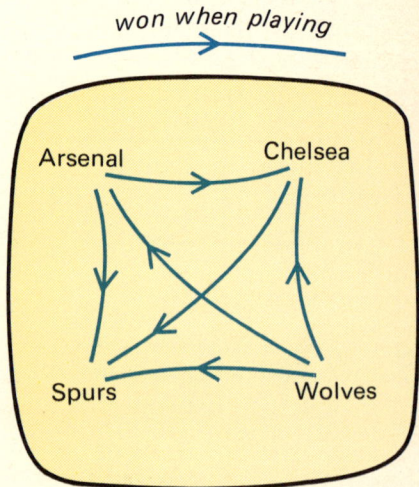

Arsenal   Chelsea

Spurs   Wolves

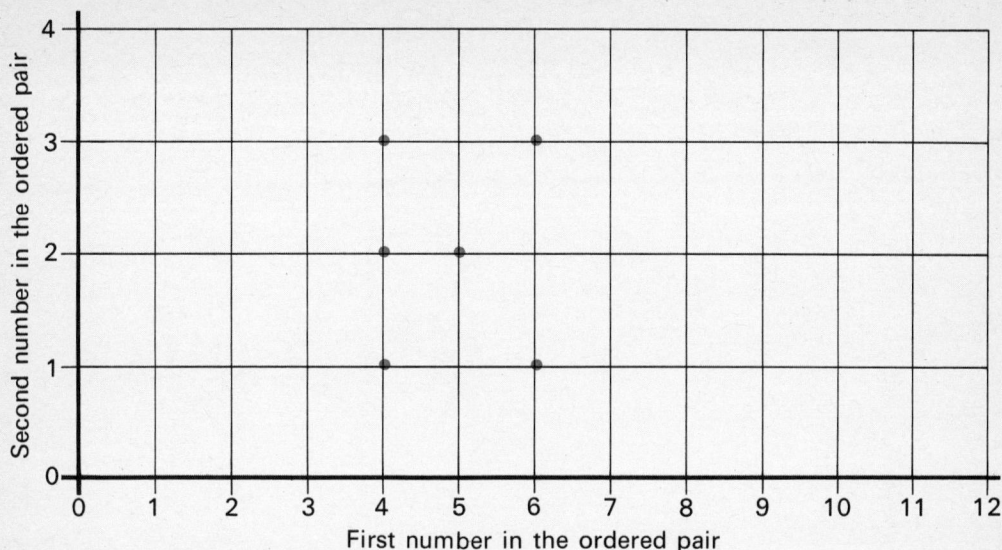

A   Copy the graph above (W/s 16).
    For each dot write down an ordered pair to show its position.

B   On your graph for A draw a straight line from:
    1. (6, 1) to (4, 1)   2. (4, 1) to (4, 2)   3. (4, 2) to (4, 3)
    4. (4, 3) to (6, 3)   5. (4, 2) to (5, 2)
    What do you notice?

C   Show on your graph for A the points: (7, 3), (7, 1), and (9, 1).
    Join (7, 3) to (7, 1) and (7, 1) to (9, 1).
    What letter have you drawn?

D   Show on your graph for A the points: (10, 1), (10, 2), (10, 3),
    (12, 3), and (12, 2).
    Join (10, 1) to (10, 2), (10, 2) to (10, 3), (10, 3) to (12, 3),
    (12, 3) to (12, 2), and (12, 2) back to (10, 2).
    What letter have you drawn this time?

E   What letter could you put in front of the three letters on your
    graph to make a word?
    Write down the set of ordered pairs for this letter, and also
    the instructions for joining the points.
    Ask a friend to use your ordered pairs and instructions. Does
    he get your word?

(2,8)   (5,20)

(3,12)   (8,32)

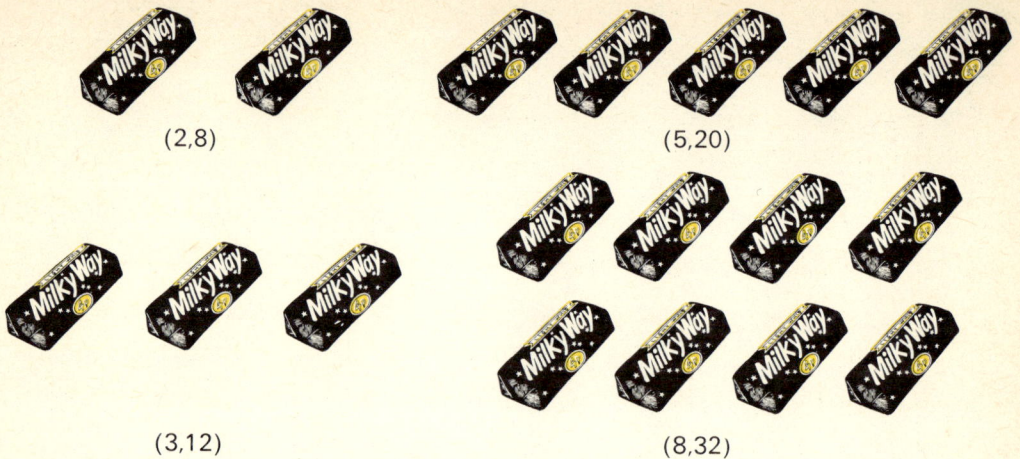

A   Look at the diagram above.
     What information is shown by the ordered pairs?
     What does the first number in each ordered pair represent?
     What does the second number in each ordered pair represent?
     Write down some more ordered pairs which belong to this set.

B   Write down a statement about the two numbers in each pair in A.

C   If the first number in a pair in A is ☐, what is the second?

D   Using your ordered pairs
     in A, copy and complete
     the graph on the right
     (W/s 16).

E   Use your graph for D to
     find the cost of six
     bars of chocolate.

F   Use your graph for D to
     find how many bars of
     chocolate can be bought
     for 36p.

G   What can you say about the set of points in your graph for D?
     Why would it be incorrect to join these points with a line?

H   If the first number in a pair in A is n, what is the second?

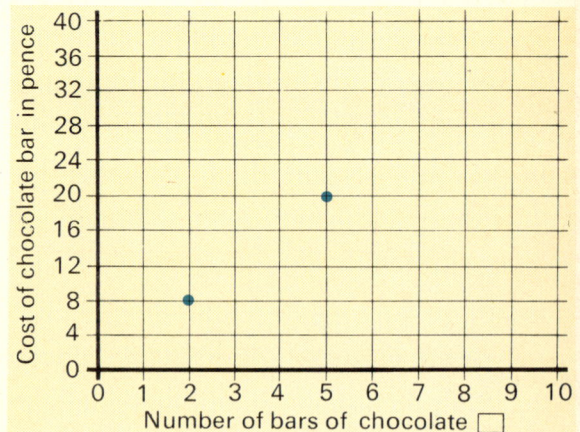

A   Find some more ordered pairs
    for the set shown on the right.

B   Write down a statement about the
    two numbers in each pair in A.

C   If the first number in a pair
    in A is ☐, what is the second?

D   Draw a graph to show your set of
    ordered pairs in A.
    What can you say about your graph?

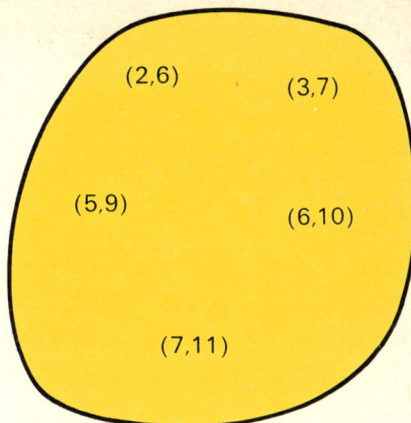

E   If ☐ stands for the first number in an ordered pair, write
    down the second number:
    1. {(0, 6), (1, 7), (2, 8), (3, 9), (4, 10), …}
    2. {(1, 5), (3, 15), (4, 20), (6, 30), (7, 35), …}

F   Look at the set of points below. Write down the co-ordinates of
    each point. If ☐ stands for the first number in a pair, write
    down the second number.

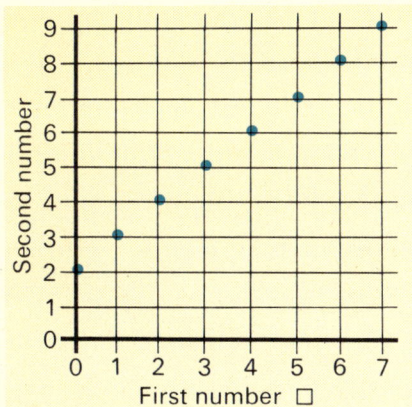

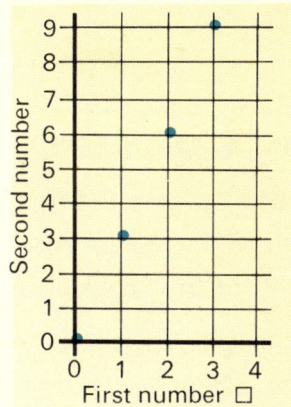

G   What can you say about the graph of each set of points above?

H   Write down three or four pairs belonging to the set described by:
    1.(☐, ☐ + 7)   2. (☐, ☐ − 6)   3. (☐, 7 − ☐)   4. (☐, ☐ × 6)

I   For each set of ordered pairs in H, draw a graph (W/s 17, 18).
    In what ways are the graphs for H1 and H2 similar?
    In what ways does the graph for H4 differ from those for H1 and H2?

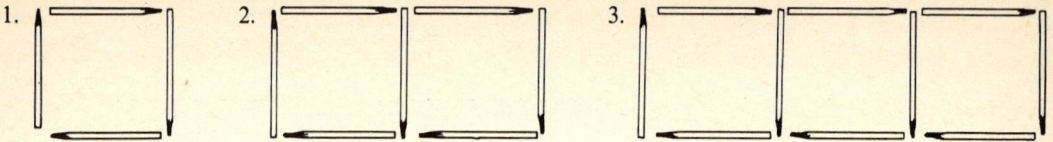

1.   2.   3.

A   Look at the three shapes above which are made from matchsticks. How many matchsticks are there in each of the three shapes?

B   How many matchsticks will there be in the fourth shape in the set? How many matchsticks will there be in the fifth shape? Draw these two shapes to check your answers. (One will have four squares, the other five squares.)

C   Use your answers to A and B to complete the table below:

Number of squares in shape	1	2	3	4	5	6
Number of matchsticks in shape	4					

D   How many matchsticks will there be in the tenth shape? Say in your own words how you worked out your answer.

E   Copy and complete the graph on the right (W/s 19) using the ordered pairs you found in C. The first, (1, 4), is shown. What can you say about the set of points on your graph?

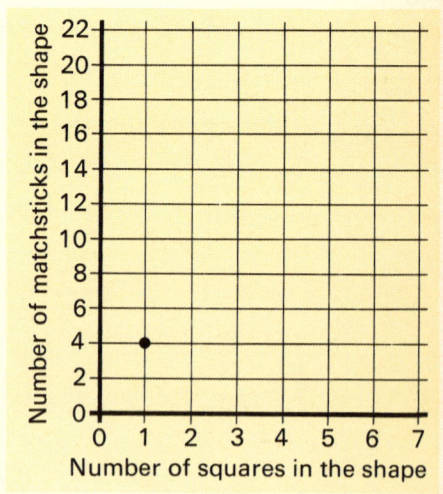

F   If ☐ represents the first number in each ordered pair in C, write down the second.

G   How many matchsticks will there be in the nth shape?

H   Look at the shape on the right. How many rectangles can you find in this shape? Investigate the number of rectangles in each of your matchstick shapes in B.

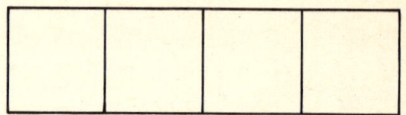

A   Copy and complete the arrow graph

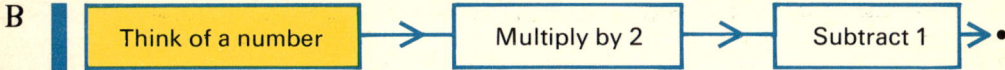

$7 \xrightarrow{\times 2} \bullet \xrightarrow{-1} \bullet$

B

| Think of a number | → | Multiply by 2 | → | Subtract 1 | → • |

Use this flow chart for four or five numbers.
Record each result using the ordered pair:
(number thought of, final answer).
What can you say about the number you finish with in each case?

C   Copy and complete the arrow graph:

1. $5 \xrightarrow{+3} \bullet \xrightarrow{+2} \bullet$

2. $10 \xrightarrow{\times 7} \bullet \xrightarrow{-3} \bullet$

3. $\square \xrightarrow{+3} \bullet$

4. $\square \xrightarrow{\times 5} \bullet$

5. $\square \xrightarrow{\times 2} \bullet \xrightarrow{-1} \bullet$

6. $\square \xrightarrow{\times 5} \bullet \xrightarrow{+3} \bullet$

D   Use the set of numbers {1, 2, 3, 4, 5, 6} in the flow chart below.

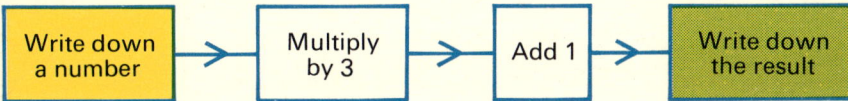

| Write down a number | → | Multiply by 3 | → | Add 1 | → | Write down the result |

What do you notice about the numbers you start and finish with?
How many more is each result than the previous result?

E   Use the set of numbers {1, 2, 3, 4, 5, 6} in the flow chart below.

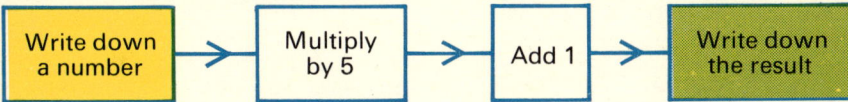

| Write down a number | → | Multiply by 5 | → | Add 1 | → | Write down the result |

What do you notice about the numbers you start and finish with?
How many more is each result than the previous result?

F   If you had used the number n in the flow chart in D,
what would you have written down as your result?

G   If you had used the number n in the flow chart in E,
what would you have written down as your result?

H   If your result had been 46, what would have been your
original number in: 1. D?   2. E?

Representing information

FOOTBALL RESULTS

Division One

ARSENAL 1	0...SOUTHMPTN	
Ball	19,210	
COVENTRY.... 0	1....BIRMINGHM	
	Hatton 27,825	
DERBY..........1	1...........STOKE	
Bourne	Ritchie 28,176	
IPSWICH......1	1..........NORWICH	
Hamilton	Boyer 25,004	
LEEDS 1	1.....NEWCASTLE	
Clark	Barrowclough	
	46,611	
LEICESTER 2	1........EVERTON	
Workington	Latchford	
Earle	22,286	
LIVERPOOL ...1	0.........BURNLEY	
Toshack	42,562	
Q.P.R 3	1....TOTTENHAM	
Givens,Bowles	Chivers	
Francis	25,775	
SHEFF.UTD....0	1.......MAN.UTD	
	Macari 29,203	
WEST HAM ... 3	0........CHELSEA	
Bonds 3	34,043	

Division Two

ASTON V.......1	3.....WEST BROM	
Morgan	Wile Brown 2	
	37,323	
BLACKPOOL...0	2 ....BOLTON	
	G Jones Brown	
	18,575	
BRISTOL C ... 0	0........OXFORD U	
	25,438	

CARLISLE 2	2........ PRESTON
O'Neill,Owen	Elwiss 27,671
HULL............. 2	1....... SHEFF WED
Greenwood.	Potts 8,193
Pearson	
PORTSMTH 0	0............ MILLWALL
	11,004
SUNDERLND ...0	2 .... MIDDLSBRO
41,658	Mills, Foggon
SWINDON......1	1.............. CARDIFF
Compton	Powell 5,126
C.PALACE..... 0	0 ....FULHAM
	11,761
NOTTS CO 1	1................. LUTON
Randall	Ryan 15,613
ORIENT...........1	1....... NOTTM FOR
Bullock	Jackson 10,503

Division Three

BLACKBURN.. 0	1............ OLDHAM
	Groves 10,244
BOURNEM'TH 3	2 .... HEREFORD
Sainty	Radford 8,370
Chadwick	Naylor
Greenhalgh	
GRIMSBY 1	1..................YORK
Fletcher	Seal Lyons 7,410
HALIFAX..... 0	0...... HUDDRSFLD
	8,126
PLYMOUTH 1	0..............BRISTOL
Provan	11,373
PORT VALE 2	1.........SOUTHPORT
O'Neill (o.g)	Russell 3,200
Gough	
ROCHDALE....1	2........CHESTRFLD
Brennan	Bellamy Wilson
	1,546
WATFORD..... 1	0...... SHRFWSBRY
Scullion (pen)	5,860

WREXHAM 0	0........ TRANMERE
	7,548
WALSALL..... 0	0.....CHARLTON A
	4,781
BRIGHTON.... 0	2........ALDERSHOT
	Dean 2 6,284
SOUTHEND 1	1....... CAMBRIDGE
Brace	Greenhalgh 4,852

Division Four

COLCHESTER ..0	1.....GILLINGHAM
	Yeo 8,082
PETERBORO ... 2	0..............BURY
Cozens, Robson	9,641
TORQUAY.......1	1..... NORTHMPTN
Twitchin	Stratford 3,715
NEWPORT C ...1	0............READING
D.Jones	2,167
SWANSEA 1	0.... BRADFORD C
Davies	4,093
LINCOLN C..... 3	3......... EXETER C
Spencer,Ward	Binney 2 5,928
Smith	Morrin
BARNSLEY..... 3	1.......... CHESTER
Butler 2	Mason 2,385
Greenwood	
MANSFIELD ...1	0..........HARTLEPL
McCaffrey	4,331
CREWE...........1	1.....DARLINGTON
Morritt (o.g)	Atkins 1,874
ROTHERHM....0	3...........BENTFORD
	Cross 3 4,871
SCUNTHRPE ...0	0.....WORKINGTON
	10,719
STOCKPORT...0	3 .....DONCASTER
	Kitchen Elwiss
	O'Callaghan 1,624

A   On a grid (W/s 19), like that shown on the right, draw a scattergram to show the results of all the matches given above.

B   From your scattergram find:
1. the result which occurred most often
2. the number of matches in which 4 goals were scored
3. the number of matches in which less than 2 goals were scored.

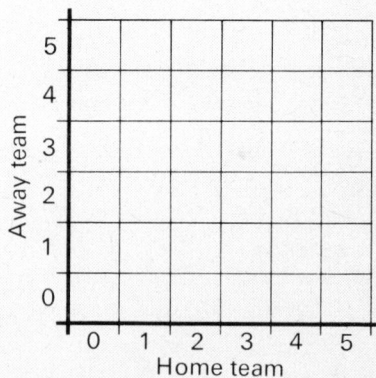

C   Draw a scattergram to show the results of matches on a recent Saturday. Write down all you can find out from your scattergram.

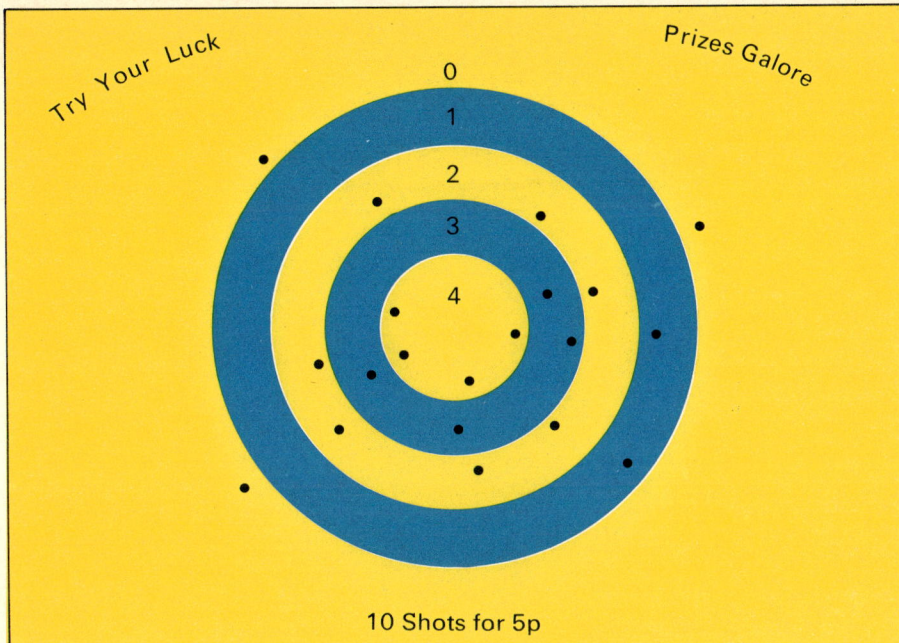

A   Terry is keen on shooting and when the fair came to his town
he tried his skill.
He had 20 shots and his card is shown above.
What was his total score?

B   Draw a block graph to show the results (W/s 20).
What score did Terry get most often?
Find Terry's average (mean) score for his shots.

C   Terry had 20 more shots and this time had a score of 60.
Make a copy of the card (W/s 20) and show how he could have
made this score. What score did Terry get most often on your
card? Find Terry's average (mean) score for his shots.
Compare your two answers with a friend. Do you agree in both
cases?

D   Use another copy of the card (W/s 21).
Close your eyes and, with a pencil, make 20 dots on it, trying to get
as near the centre as possible.
Draw a graph to show your results (W/s 21). (*Note* Some of your
dots may be on the circles themselves. In this case you must decide
upon a rule for dealing with them.) Find your average (mean) score.

A   Try this counting activity.
     Each member of the class quietly counts *one, two, three, four, five, ...*
     After exactly one minute they are stopped.
     They then write down the number reached in the minute.
     The numbers reached by all the children are then listed and the
     results are shown by a graph (W/s 22).

B   Did you find that the drawing of the graph in A took a long
     time? Could you think of any ways of simplifying the drawing?

     The graph on the opposite page shows one way in which the results
     were shown by a class.

C   Look at the graph.
     Why do you think the *numbers reached* were grouped in fives?
     Why do you think the groups started at 80, instead of 0?

D   From the graph find how many children counted to one of the
     numbers:
     1. 80, 81, 82, 83, 84   2. 85, 86, 87, 88, 89   3. 100, 101, 102, 103, 104
     4. 130, 131, 132, 133, 134   5. 105 to 109   6. 135 to 139.

E   Can you say, from the graph, whether any child finished his
     counting at:
     1. 94?   2. 110?   3. 128?   4. 131?

F   How many children were in the class?

     There were most children for the group 110 to 114.
     This group is the mode.

G   Arrange your results from A in suitable groups (all of the same size)
     and show them as a graph (W/s 23).
     Which of your groups is the mode?

H   You do not know any of the actual numbers reached by the
     children whose results are shown in the graph on the opposite
     page.
     Can you think of a way in which you could find an
     approximate value of the mean of the numbers reached?
     Find, as simply as you can, the mean of the numbers reached
     by the children in your class.

A graph to show our counting

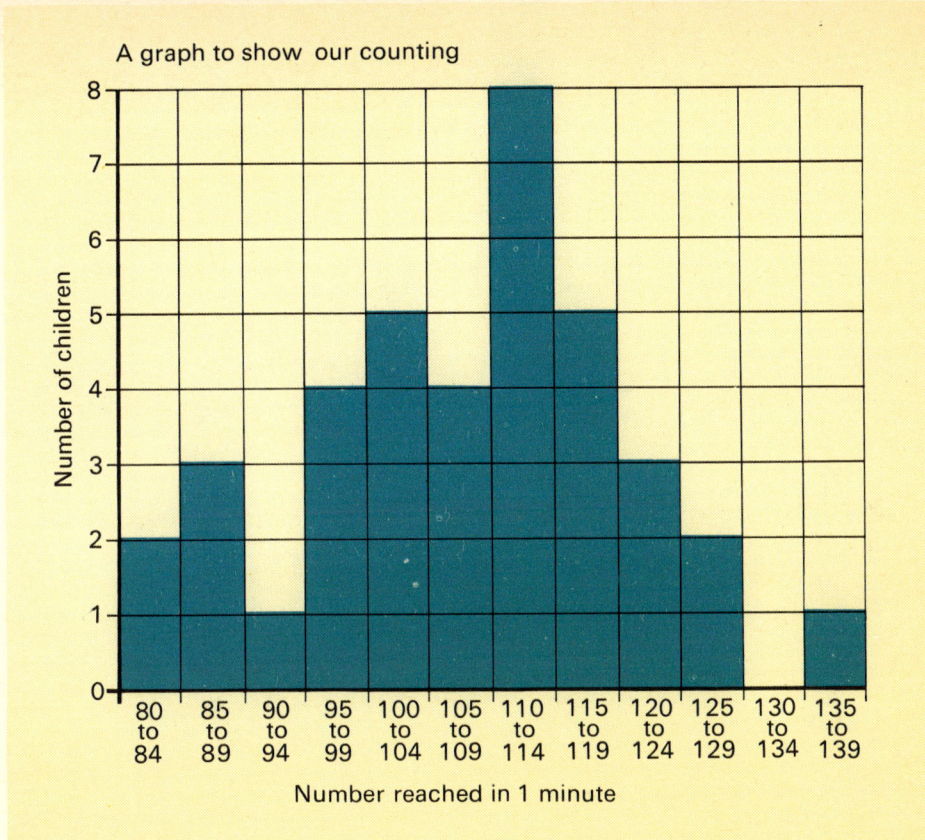

A graph to show our counting
Number of children (vertical axis)
Number reached in 1 minute (horizontal axis)

A   In another counting activity each child in a class *writes*
         1, 2, 3, 4, 5, 6, 7, 8, 9, 10, 11, 12, 13, 14, ….
After exactly one minute they are stopped.
Try this activity with your class.
Draw a graph to show your results (W/s 24).
In what ways is this graph like your graph for G on page 46?
In what ways is it different?

B   Here is another activity.
Each child in your class throws a ball as far as he can.
Measure, to the nearest metre, the length of each throw.
Draw a graph to show your results (W/s 25).

C   What is the mode for A and for B?
Find an approximate value of the mean for A and for B.

The table and graphs below show on which days of the week
a class of 36 children have their birthdays this year.

Monday	////	4
Tuesday	̶H̶H̶T	5
Wednesday	//	2
Thursday	̶H̶H̶T //	7
Friday	̶H̶H̶T ̶H̶H̶T //	12
Saturday	/	1
Sunday	̶H̶H̶T	5
		36

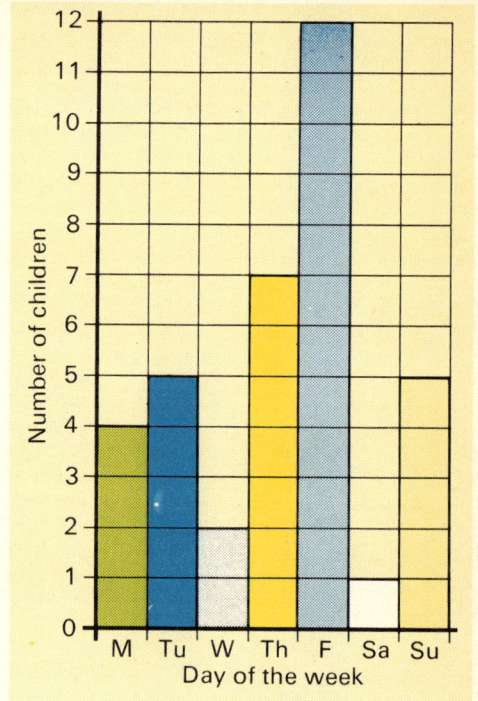

A   Look at the table above.
A tally mark was made as each
child called out the day for
his birthday.
Why do you think the fifth tally
mark is put across the first
four tally marks?

B   Look at the block graph.
Check that it is drawn correctly.

For the graph on the right a
circle is used. This kind of
graph is called a **pie chart**.

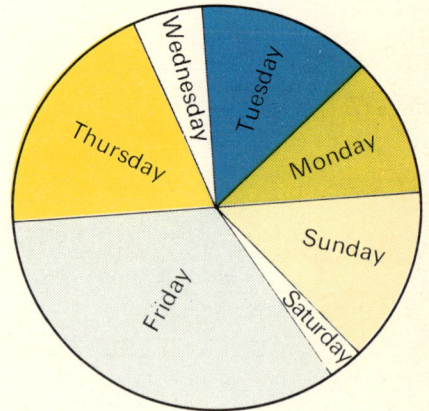

C   Why do you think the name
pie chart is used?

D   Look at the pie chart.
How do you think the size of
each region is worked out?

E   What angle at the centre of
the circle is used to show one child?
Measure the angle at the centre of the circle for each region.
Are they what you think they should be?

F   Ask 36 children about the day on which they have their birthday
this year. Show your results in three ways, as above (W/s 26).

A In a class of 32 children there are 20 boys and 12 girls.
1. What fraction of the class are: (a) boys?   (b) girls?
2. What percentage of the class are: (a) boys?   (b) girls?
3. Draw a pie chart to show the information.

Here is one way of finding the answers to A.

Fraction of the class who are boys is $\frac{20}{32}$ (i.e. $\frac{5}{8}$).

Fraction of the class who are girls is $\frac{12}{32}$ (i.e. $\frac{3}{8}$).

$\frac{1}{8}$ of $100 = 12\frac{1}{2}$

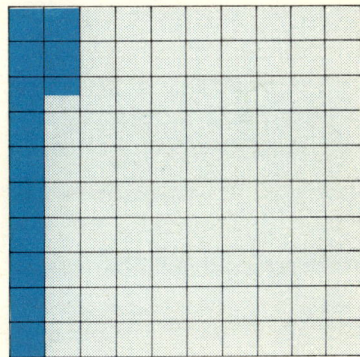

Think of the whole class as 100%.
The percentage for the boys
is $\frac{5}{8}$ of 100%.
$\frac{1}{8}$ of $100 = 12\frac{1}{2}$ $\frac{5}{8}$ of $100 = 62\frac{1}{2}$
Percentage for boys is $62\frac{1}{2}\%$
Percentage for girls
is $37\frac{1}{2}\%$ ($\frac{3}{8}$ of 100%)

The angles for the pie chart can be found as shown below.

$\frac{1}{8}$ of $360° = 45°$

$\frac{5}{8}$ of $360° = 225°$

$\frac{3}{8}$ of $360° = 135°$

12 girls
$\frac{3}{8}$ of the class
$37\frac{1}{2}\%$

20 boys
$\frac{5}{8}$ of the class
$62\frac{1}{2}\%$

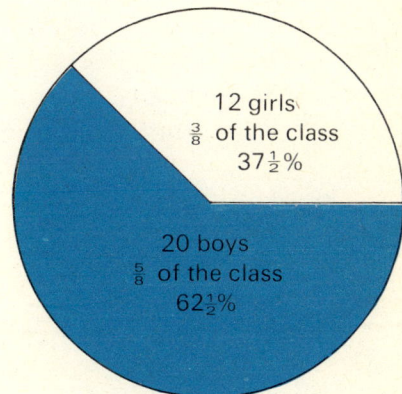

B In a class of 34 children there are 20 girls and 14 boys.
1. What fraction of the class are: (a) girls?   (b) boys?
2. What percentage of the class are: (a) girls?   (b) boys?
3. Draw a pie chart to show the information.

In B on page 49, the fractions for the girls and boys are $\frac{10}{17}$ and $\frac{7}{17}$.

The changing of fractions such as $\frac{10}{17}$ and $\frac{7}{17}$ to
percentages needs careful thought. One way is shown below.

$\frac{1}{17}$ of $100 = 5 \cdot 88 \ldots$

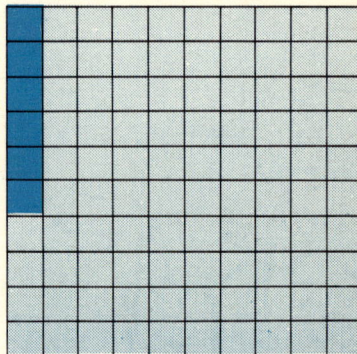

$$1 \text{ whole} = 100\%$$
$$\frac{1}{17} \quad\quad = \left(\frac{100}{17}\right)\%$$
$$= 5 \cdot 88 \ldots \%$$

$$\frac{10}{17} = \left(\frac{100}{17} \times 10\right)\%$$
$$= \left(\frac{1000}{17}\right)\%$$
$$= 58 \cdot 823 \ldots \%$$
$$= 58 \cdot 8 \ldots \% \text{ (to one place)}$$

In the same way
$$\frac{7}{17} = \left(\frac{100}{17} \times 7\right)\% = \left(\frac{700}{17}\right)$$
$$= 41 \cdot 2\% \text{ (to one place)}$$

Another way is to change each fraction to a decimal straightaway.

$$\frac{10}{17} = 10 \div 17 = 0 \cdot 58823 \ldots = 58 \cdot 823 \ldots \text{ hundredths}$$
$$= 58 \cdot 8\% \text{ (to one place)}$$

$$\frac{7}{17} = 7 \div 17 = 0 \cdot 41176 \ldots = 41 \cdot 176 \ldots \text{ hundredths}$$
$$= 41 \cdot 2\% \text{ (to one place)}$$

The finding of the two angles for the pie chart can be done as
in the example below.

$$\frac{1}{17} \text{ of } 360° = \left(\frac{360}{17}\right)°$$
$$\frac{10}{17} \text{ of } 360° = \left(\frac{360}{17} \times 10\right)° = \left(\frac{360 \times 10}{17}\right)° = \left(\frac{3600}{17}\right)° = 212°$$
$$\text{(to the nearest degree)}$$

A Change to a percentage (correct to one place of decimals):

1. $\frac{7}{9}$ 2. $\frac{5}{7}$ 3. $\frac{8}{13}$ 4. $\frac{10}{19}$ 5. $\frac{13}{21}$ 6. $\frac{14}{25}$ 7. $\frac{21}{29}$ 8. $\frac{1}{27}$

B Find the fraction of $360°$:   1. $\frac{2}{5}$ 2. $\frac{5}{9}$ 3. $\frac{7}{12}$ 4. $\frac{8}{13}$ 5. $\frac{13}{21}$

Cars	27
Bicycles	8
Vans	12
Buses	5
Lorries	16
Motor cycles	7
Scooters	2

A The results of a traffic count are shown in the table above.
1. Draw a block graph to show the results (W/s 27).
2. Draw a pie chart to show the results (W/s 27).
3. Show the numbers for each type of vehicle as a percentage of the total, to the nearest whole number (W/s 27).

B Carry out a traffic survey of your own.
Show the results (W/s 28):   1. in a table   2. as percentages
3. by a block graph   4. by a pie chart

C Collect the registration numbers of 50, or more, cars.
Sort them according to the letter which gives their year of registration. (Put those without a letter in one subset—*others*.)
Show your results (W/s 29) as: 1. a block graph   2. a pie chart

D Using an AA or RAC year book, find the place of registration of each of your cars in C. On a large map of the British Isles put a flag or drawing pin for each car at its place of registration. Write down all that you notice.

Review 1A

A   Work through the flow chart on
the right.
What is the average (mean) of the
odd numbers less than 20?

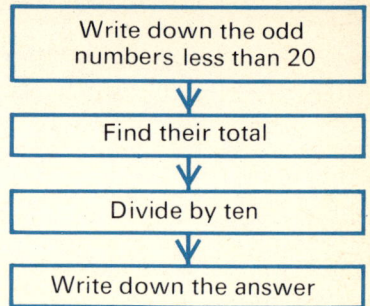

Write down the odd numbers less than 20
↓
Find their total
↓
Divide by ten
↓
Write down the answer

B   Draw and give the name of a quadrilateral with:
  1. each of its angles a right angle
  2. all its edges the same length
  3. all its edges the same length and each angle a right angle
  4. its opposite edges parallel

C   Copy and complete the Venn diagram (W/s 30).

Whole numbers from 1 to 40

Multiples of 5   Multiples of 3

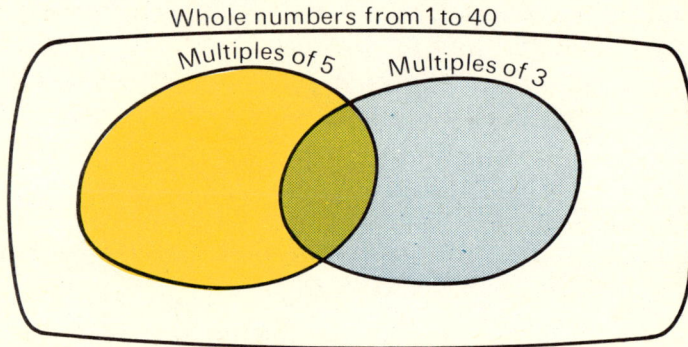

What can you say about the numbers in the intersection of the
two subsets?
What is the least number which is a multiple of 5 and of 3?

D   The heights of ten boys are: 158 cm, 159 cm, 160 cm, 160 cm,
160 cm, 162 cm, 163 cm, 163 cm, 168 cm, 170 cm.
Find the average (mean) height of the set of boys.
Do you think that the average height of the boys is exactly
what you found? If not, give reasons.

E   Give the answer to the first place of decimals:
  1. $521 \div 8$ 2. $239 \div 9$ 3. $640 \div 7$ 4. $805 \div 6$ 5. $209 \div 10$

F   Find the area of the polygon
with the dimensions shown.

G   The polygon is the base of a
prism of height 8 cm.
Find the volume of the prism.

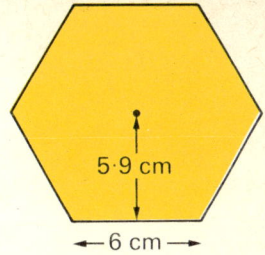

5·9 cm

◄— 6 cm —►

H   *The wind was a torrent of darkness among the gusty trees,*
The moon was a ghostly galleon tossed upon cloudy seas,
The road was a ribbon of moonlight over the purple moor,
And the highwayman came riding—
        Riding—riding—
The highwayman came riding, up to the old inn door.

Look at the first verse, above, of the poem 'The Highwayman'.
Count the number of letters in each word.
Find how many words have: 1 letter, 2 letters, 3 letters, etc.
Show your result by using the ordered pair:
        (number of letters in a word, number of words).
Also show your results by drawing a graph (W/s 31).

I   The graph shows the colours
of the dresses worn by 20 girls.
1. What angle at the centre
of the circle is used to
show one girl?
2. Measure each of the four
angles at the centre of the
circle.
Use your results to find
how many girls wore a
dress of each colour.
3. Write down the fraction
of the girls wearing each
colour.
4. Find the percentage of the
girls wearing each colour.

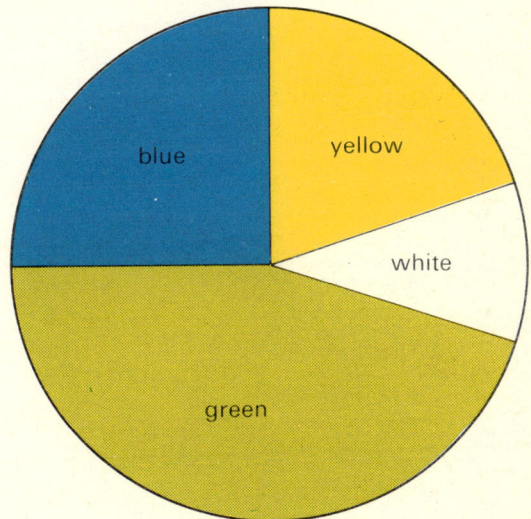

blue

yellow

white

green

J   Find the area of a rectangle whose dimensions are:
1. length 12·6 cm, width 9 cm   2. length 24·7 m, width 18 m
3. length 42·3 cm, width 8 cm   4. length 82·4 m, width 57 m

Review 1B

A Follow the flow chart for:
1. $\{1, 2, 3, 4, 5, 6\}$
2. $\{2, 4, 6, 8, 10, 12, 14\}$
3. $\{5, 10, 15, 20, 25, 30, 35, 40\}$
4. $\{\frac{1}{2}, 1, 1\frac{1}{2}, 2, 2\frac{1}{2}, 3, 3\frac{1}{2}, 4\}$

B Find the sum of the members of each set in A.

C Compare your results for A and B. What do you notice?

D Try using the flow chart to find the sum of the whole numbers from 1 to 100 (both inclusive).

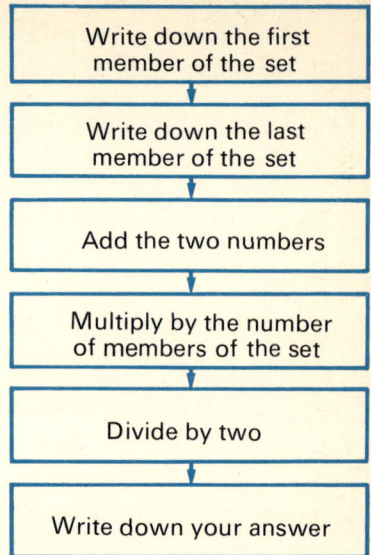

> Write down the first member of the set
>
> ↓
>
> Write down the last member of the set
>
> ↓
>
> Add the two numbers
>
> ↓
>
> Multiply by the number of members of the set
>
> ↓
>
> Divide by two
>
> ↓
>
> Write down your answer

E Draw a triangle whose edge lengths are 5·7 cm, 8·2 cm, and 10·8 cm. Find the area of the triangle.

F Complete the statement. Show your result on a Venn diagram.
1. $\{1, 4, 9, 16, 25, 36, 49, 64, 81, 100\} \cap \{1, 8, 27, 64, 125\} =$
2. $\{T, R, I, A, N, G, L, E\} \cap \{T, R, A, P, E, Z, I, U, M\} =$
3. $\{\triangle \ \square \ \pentagon \ \hexagon \ \bigcirc\} \cap \{\triangle \ \square \ \rhombus \ \ominus\} =$

G What percentage of the numbers *between* 1 and 40 are prime numbers?

H Copy and complete the table about cuboids (W/s 30).

Length	6 cm	12 cm	7 cm	8 cm	
Width	4 cm	4·5 cm	4 cm		2·5 cm
Height	5 cm	2·3 cm		6 cm	2·5 cm
Total surface area					
Volume			140 cm³	259·2 cm³	50 cm³

I Using $\{0, 1, 2, 3, 4, 5\}$ for $\square$, write down the set of ordered pairs: 1. $(\square, \square + 1)$ 2. $(\square, \square + 2)$ 3. $(\square, \square + 3)$
4. $(\square, \square)$
On the same graph show each set of ordered pairs (W/s 32).
What do you notice?
What can you say about the graph of the ordered pairs for $(\square, \square + 10)$?

Review 1C

A | Find the number which each letter stands for:

1.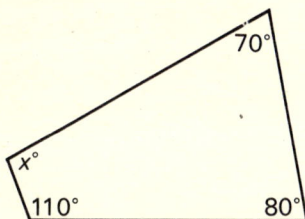

2.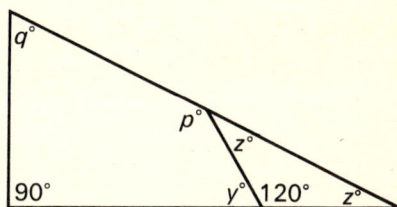

B | If X = {S, P, A, I, N},   Y = {F, R, A, N, C, E}, and
Z = {P, O, R, T, U, G, A, L} write down the set:
1. $X \cap Y$ 2. $X \cap Z$ 3. $Y \cap Z$ 4. $X \cap Y \cap Z$.

C | Write down the mean of:
1. p, q, r, s, and t 2. $3a$, $3b$, and $3c$ 3. $p^2 + q^2$.

D | Copy and complete the table, about cuboids (W/s 32).

Length Width Height	l cm w cm h cm	a cm b cm	p cm q m r cm	d cm d cm e cm	n cm n cm n cm
Surface area					
Volume		V cm³			

E | Using {1, 4, 7, 10, 13, ...} for ☐, write down the first three
ordered pairs of the set represented by (☐, ☐ × 4).
Also write down the 10th ordered pair and the nth ordered pair.

F | Using {0, 1, 2, 3, 4, 5, 6} for ☐, draw a graph (W/s 32) to show the
ordered pairs (☐, ☐ × n) for n = 1, 2, 3, and 4. What do you
notice?

G | Make a cube from plasticine or potato.
Cut off one corner as shown in the diagram
on the right. You should have made
a triangular face. How would you
have to cut the cube to make:
1. a rectangle?   2. a parallelogram?

What other shapes can you make
with one cut? Record your results.

Measures

Peter and his friends helped Mr. Evans, the farmer, to harvest his potatoes. Peter was very interested in seeing the sacks being filled and stacked. He wondered what the total mass of the potatoes would be. He asked Mr. Evans.

Mr. Evans said, 'I expect to get about 30 tonnes of potatoes per hectare, and I'm using about 16 hectares of land, so you can work out what the total will be.' Peter found the answer (480 tonnes) quite easily but was a little puzzled about **tonnes** and **hectares**.

A   Find out all you can about tonnes and hectares.
Look for references to tonnes on trucks, lorries, etc.
Look for references to hectares in Estate Agents' windows and in advertisements, etc.
Record all you find out.

B   What is an **are**? What is the relationship between a hectare and an are?

C   Mark out an are on the school playing field.
If possible, also mark out a hectare.

D   Estimate the area of a netball pitch, a hockey pitch, a tennis court, etc. in hectares or in ares.
Find the area of each. Compare your estimates and measurements.

A Follow the flow chart below, for each sports pitch shown on the right.

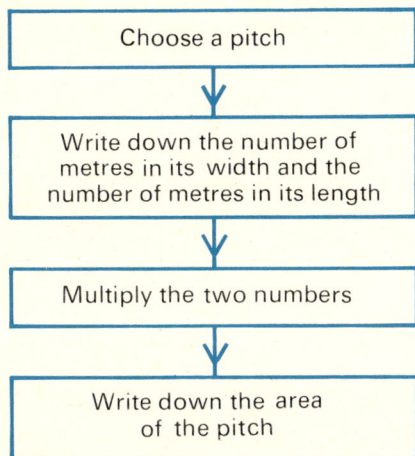

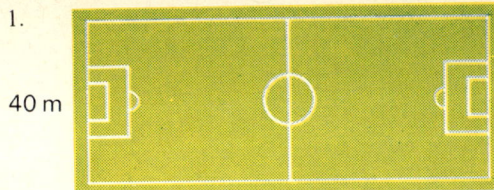

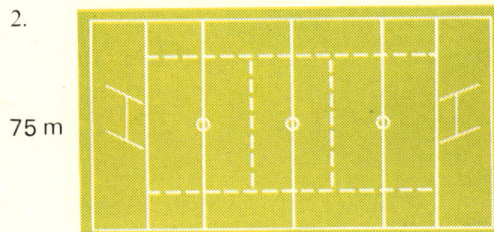

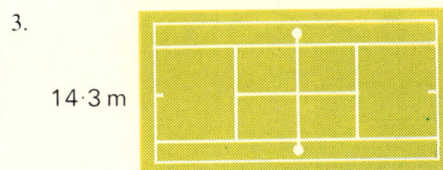

1.

40 m

100 m

Choose a pitch

↓

Write down the number of metres in its width and the number of metres in its length

2.

75 m

140 m

↓

Multiply the two numbers

↓

Write down the area of the pitch

3.

14·3 m

29 m

B Write down the area of each sports pitch in A in:
(a) ares (b) hectares.

Program

C Work through the program shown on the right.
What is the number output in instruction ④?
Can you use this program to find the area of a rectangle?

①   5 → **W**
②   12 → **L**
③   **W** × **L** → **A**
④   OUTPUT **A**
⑤   STOP

D How would you have to change the program in C to find the area of a rectangle with width 16·2 cm and length 23·5 cm?
What is the area of this rectangle?

E Using letters, the width of a rectangle can be written as w cm, and the length as l cm.
Using these letters, copy and complete the flow chart on the right to show how to calculate the area of a rectangle.

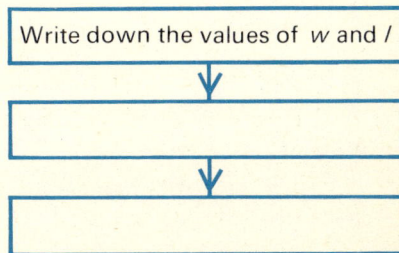

Write down the values of w and l

↓

↓

12 cm

2 cm

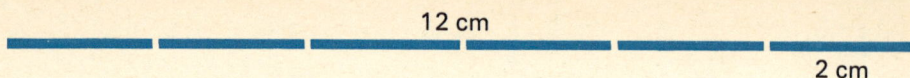

The drawing shows a strip of metal cut into 2 cm strips.
The number of 2 cm strips is given by 12 ÷ 2.

This small drawing shows a strip, of length 1·2 cm,   1·2 cm
divided into small strips, each of length 0·2 cm.
The number of small strips is given by 1·2 ÷ 0·2.   0·2 cm

A   Can you explain why the answers to the two divisions, 12 ÷ 2
and 1·2 ÷ 0·2, are the same?

B   In giving an answer to A, you probably used the idea of changing
both the 1·2 cm and the 0·2 cm to millimetres.
Working in millimetres, the division then became 12 ÷ 2.
Use this same idea to find the answer:

1. 1·8 ÷ 0·2   2. 2·4 ÷ 0·6   3. 1·8 ÷ 0·6   4. 0·8 ÷ 0·2
5. 2·4 ÷ 1·2   6. 3·6 ÷ 1·2   7. 8·8 ÷ 1·1   8. 9·6 ÷ 0·6
9. 2 ÷ 0·4   10. 7 ÷ 0·5   11. 12 ÷ 0·8   12. 20 ÷ 0·8

C   Copy and complete the arrow graph (W/s 33). Find the answer to
each division.

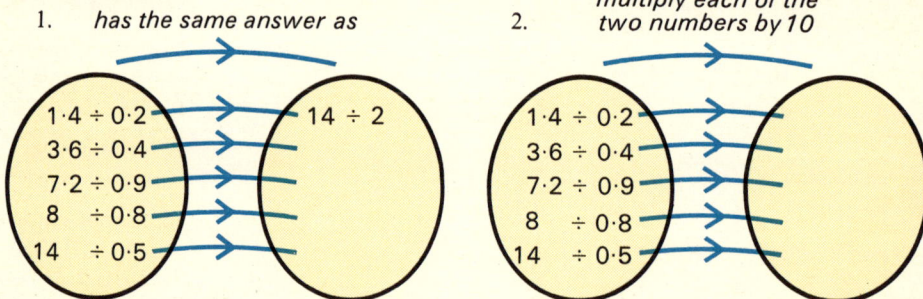

1.   *has the same answer as*   *multiply each of the*
2.   *two numbers by 10*

1·4 ÷ 0·2 → 14 ÷ 2   1·4 ÷ 0·2
3·6 ÷ 0·4   3·6 ÷ 0·4
7·2 ÷ 0·9   7·2 ÷ 0·9
8 ÷ 0·8   8 ÷ 0·8
14 ÷ 0·5   14 ÷ 0·5

What do you notice?
Explain how you would divide *any* number by a number such as 0·7.

D   1. 43·2 ÷ 1·2   2. 62·9 ÷ 1·7   3. 18 ÷ 2·4   4. 94·5 ÷ 3·5

E   Give the answer correct to one place of decimals:
1. 24 ÷ 0·7   2. 56·4 ÷ 0·9   3. 83 ÷ 1·3   4. 3·5 ÷ 0·8

F   Give the answer correct to two places of decimals:
1. 0·5 ÷ 0·9   2. 63·7 ÷ 5·3   3. 124·6 ÷ 8·7   4. 0·3 ÷ 1·3

A From a 6-metre coil of wire a machine cuts off pieces each 8 cm long. How many pieces are obtained?

B Copy and complete the table (W/s 33):

Length of coil Length of each piece	5 m 5 cm	2·4 m 6 cm	5·4 m 9 cm	8·4 m 7 cm
Number of pieces				

C Change to centimetres:

1. 0·4 m   2. 0·25 m   3. 0·67 m   4. 0·01 m   5. 0·05 m   6. 0·14 m.

D Copy and complete the table (W/s 33):

Length of coil Length of each piece	4 m 0·08 m	6 m 0·03 m	7·2 m 0·08 m	8·04 m 0·06 m
Number of pieces				

E Copy and complete the arrow graph (W/s 33). Find the answer to each division.

1.   *has the same answer as*

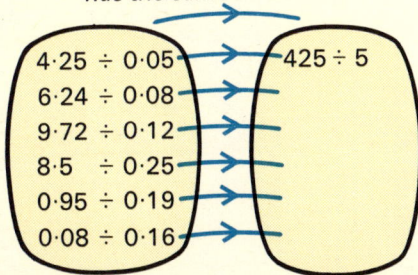

$4·25 ÷ 0·05 → 425 ÷ 5$
$6·24 ÷ 0·08$
$9·72 ÷ 0·12$
$8·5 ÷ 0·25$
$0·95 ÷ 0·19$
$0·08 ÷ 0·16$

2.   *multiply each of the two numbers by* 100

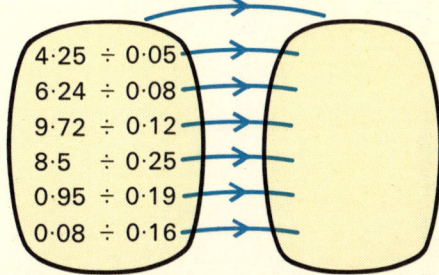

$4·25 ÷ 0·05$
$6·24 ÷ 0·08$
$9·72 ÷ 0·12$
$8·5 ÷ 0·25$
$0·95 ÷ 0·19$
$0·08 ÷ 0·16$

What do you notice? Explain how you would divide *any* number by a number such as 0·34.

A   If the area of a rectangle is 30 cm², and the length of the rectangle 6 cm, find the width of the rectangle.

B   Follow the flow chart, shown on the right, for each rectangle drawn below.

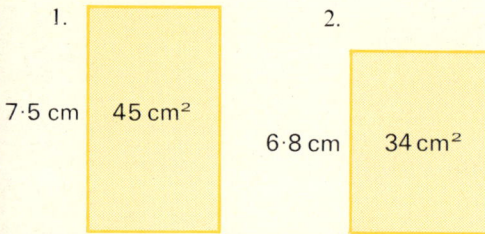

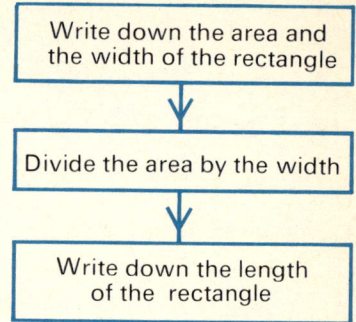

Write down the area and the width of the rectangle
↓
Divide the area by the width
↓
Write down the length of the rectangle

1.

7·5 cm   45 cm²

2.

6·8 cm   34 cm²

C   Work through the program shown on the right.
What is the number output?
Can you use this program to find the length of a rectangle when you know the area and width?

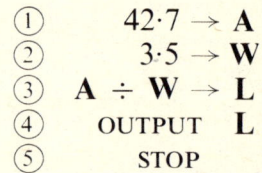

①   42·7 → **A**
②   3·5 → **W**
③   **A** ÷ **W** → **L**
④   OUTPUT **L**
⑤   STOP

D   Find the area of the triangle shown on the right.

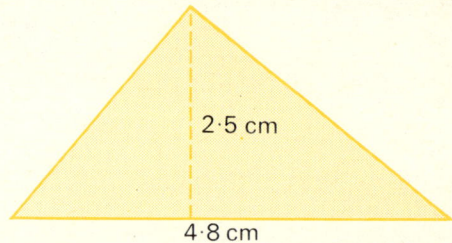

2·5 cm

4·8 cm

E   Draw a flow chart to show someone else how to find the area of a triangle.

F   Work through the program shown on the right.
The number output could be the area of a triangle.
What is:
1. the height of the triangle?
2. the length of the base of the triangle?

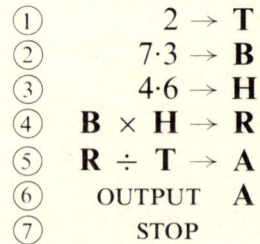

①   2 → **T**
②   7·3 → **B**
③   4·6 → **H**
④   **B** × **H** → **R**
⑤   **R** ÷ **T** → **A**
⑥   OUTPUT **A**
⑦   STOP

G   Find the height of the triangle (*not* drawn to scale).

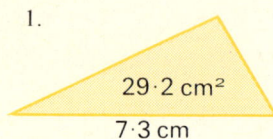

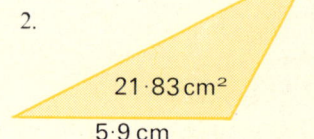

1.

29·2 cm²

7·3 cm

2.

21·83 cm²

5·9 cm

A   Find the area of each of these three triangles.
What do you notice about your results?

B   Measure the edges and the angles of each triangle. What can you say about the three triangles? In how many ways have you found the area of a triangle?

C   Draw a triangle with edges of length 5 cm, 7 cm, and 9 cm. Find its area in three ways.

D   Find the area of the trapezium shown on the right.

E   Draw another trapezium with PQ = 6 cm, RS = 3 cm, and with the same height 4 cm, but change the angle P.
Find the area of this trapezium.
What do you notice?
Why is this so?

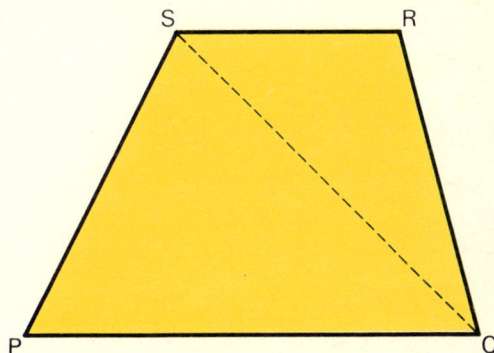

F   Find the area, A cm², of the trapezium shown on the right. Write down a formula showing the relationship between A, a, b, and h.

G   Draw a flow chart to show someone else how to find the area of a trapezium.

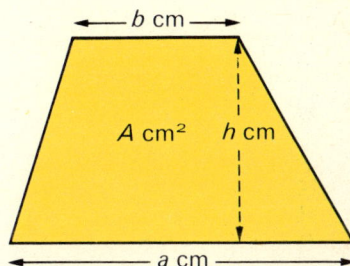

Number sequences

Mr Garfield

Mrs Garfield

Mr Davies

Mrs Davies

Mr Garfield

Mrs Garfield

Raymond Garfield

Raymond Garfield, his parents and grandparents are shown above. A simple family tree can be drawn to show Raymond's male and female ancestors. It will start like this:

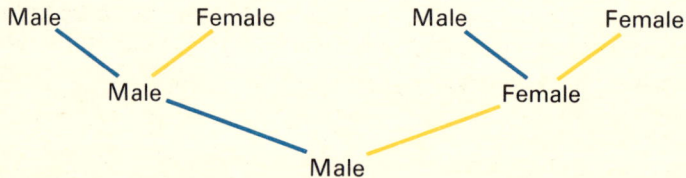

Male   Female   Male   Female

Male   Female

Male

A   On a large sheet of paper copy the family tree shown above, and extend it backwards for two more generations.

B   At the side of your tree write down the number of ancestors in each generation.
What do you notice about the numbers in this **sequence**?

C   How many ancestors did you have: 1. 8 generations ago?
2. 12 generations ago?   When do you think they were alive?

D   Do the results on this page suggest that the population of the world is decreasing? Explain your answer.

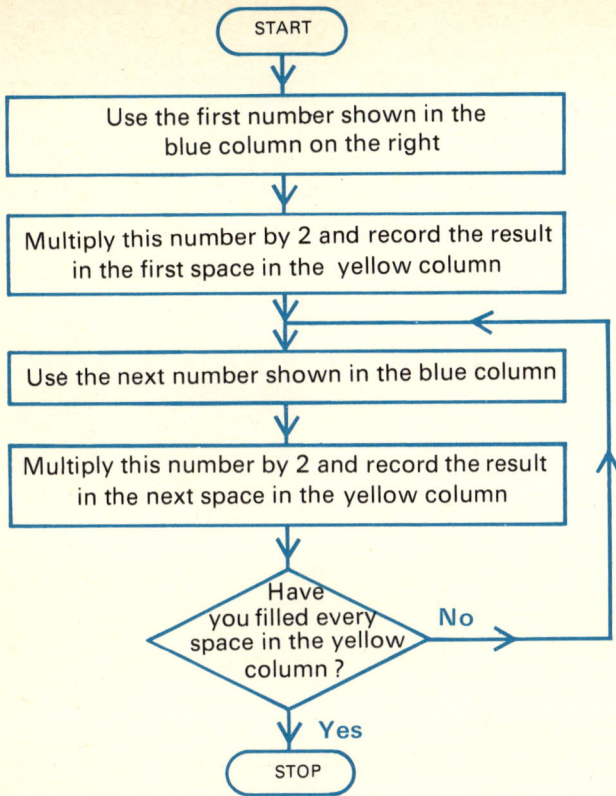

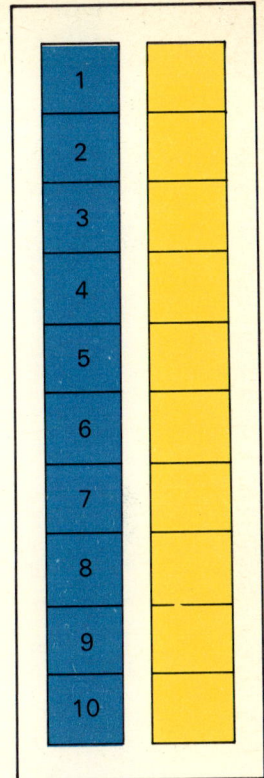

A   Carry out the instructions in the flow chart above.
    What do you notice about the numbers you have recorded?
    Write down the next two numbers which would be recorded in the
    yellow column (W/s 34).

B   ①   Cut out a long strip of paper.
    ②   Fold it in half, then unfold it.
    ③   Record the number of sections (W/s 34).
    ④   Record the number of fold lines.
    ⑤   Refold the paper.
    ⑥   Fold it in half again, then unfold it.
    ⑦   Go back to instruction ③ and repeat.

Number of folds	Number of sections	Number of fold lines
1	2	1
2		
3		
4		

C   Write down all you can about the two
    sequences of numbers you found in B.

A male bee, called a drone, has
a mother (a queen) but no father.
A queen has a mother (a queen) and
a father (a drone).

The family tree below traces a
drone's ancestors back as far
as his grandparents.

queen   drone   **2**

    queen   **1**

    drone   **1**

A   On a large sheet of paper, copy the tree
and extend it backwards for four more
generations. Assume that the drones in these
generations come from different hives.

B   Write down the number of bees in each generation. Look carefully
at the number sequence produced and write down all you find out.

C   Write down the next four numbers in the sequence.

The numbers form part of the **Fibonacci** sequence.
Leonardo Fibonacci of Pisa, who was born about 1175, discovered
the sequence when he was trying to solve a problem connected
with the breeding of rabbits.

D   The black numbers below are the first ten numbers in the
Fibonacci sequence: 1   1   2   3   5   8   13   21   34   55
                  0   1   1   2   3   5   8   13   21
The blue numbers have been produced from the black numbers. Can
you see how this was done? Write down the next two blue
numbers along the line. Can you use the blue numbers to enable
you to write down the next two black numbers in the top line?

E   Use the same approach as in D to produce another line of numbers
from the blue numbers. Write down all you find out.

F   Repeat E for: 1. 1, 2, 4, 8, 16, 32, 64, 128 ...   2. 1, 4, 9, 16, 25, 36, 49, 64 ...
What do you discover? Try this with other sequences of your choice.

START

Copy the diagram shown
on the right (W/s 34)

Add the last number in the blue
column to the last number in the
yellow column and record this
number in the blue column

Add the last number in the blue
column to the last number in the
yellow column and record this
number in the yellow column

Have
you recorded   No
ten new
numbers

Yes

STOP

B	Y
1	1
2	3

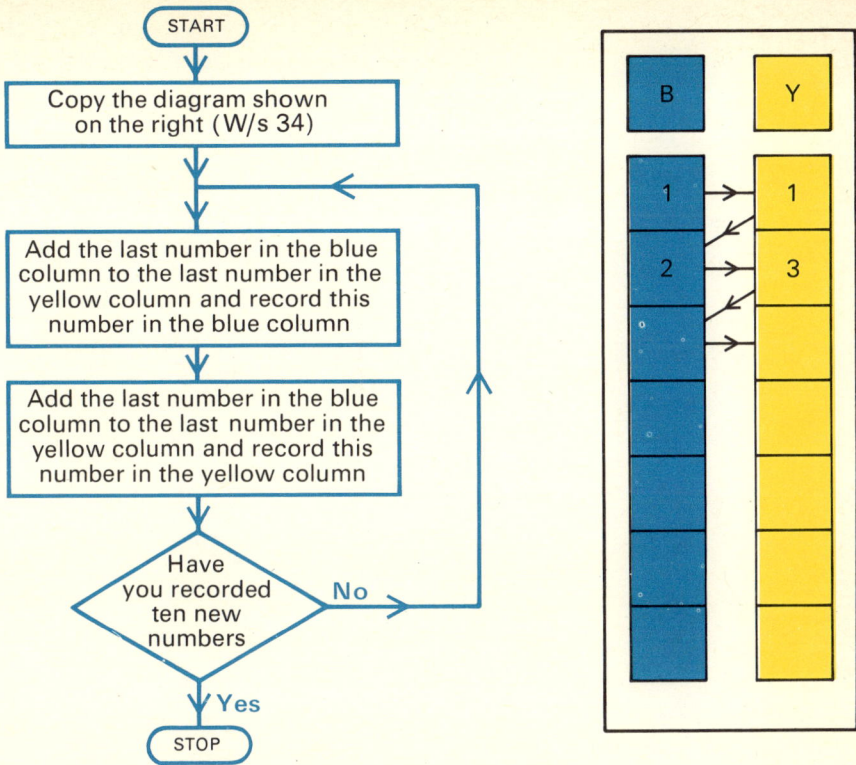

A   Carry out the instructions in the flow chart above.
What do you notice about the sequence of numbers you have
recorded?
Write down the next two numbers that would be recorded in the
yellow column.

B   Use your matchbox computer
with two boxes labelled **B**
and **Y**. Put a 1 in each.
Carry out the program
shown at the bottom of the
page. What do you notice
about the sequence of
numbers in the output?

B   Y   OUTPUT

Note: Instruction ⑤ says '*go back
to instruction* ① *and start all over again*'.

① **B + Y → B**
② OUTPUT **B**
③ **B + Y → Y**
④ OUTPUT **Y**
⑤ GO TO instruction①
⑥ STOP

C   Did you reach instruction ⑥?

INPUT   OUTPUT

2

5

7

10

A machine for producing numbers

multiply by 3   →   add 1

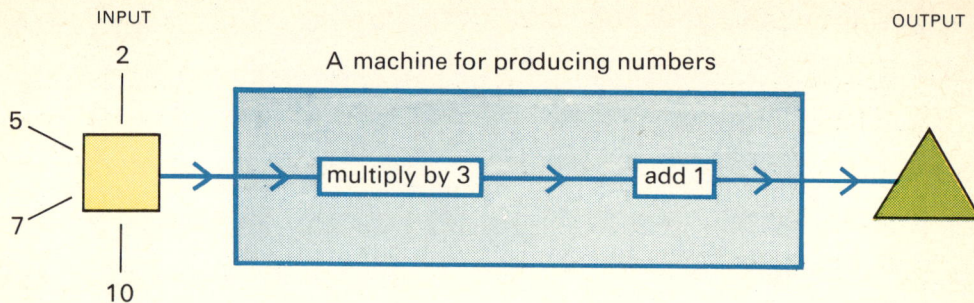

A   The machine in the diagram above will operate on any number put
     into the yellow box on the left. What number will be produced
     by the machine if you put a 2 in the yellow box?

B   Use the numbers 5, 7, and 10 in the machine. For each write down
     the number produced.

C   If the number n is put into the yellow box, what number will the
     machine produce?

D   What number would have to be put into the yellow box if the
     machine produced the number: 1. 10?   2. 13?   3. 301?   4. 1?

E   Find out what happens if the operators *multiply by 3* and *add 1*
     had been used in the opposite order.

F   If the machine at the top of the page produces the number r, what
     number would have to be put in the yellow box?

G   Work through the program   ①   1 → A
     on the right.   ②   3 → C
     Use the numbers 5, 7, 10   →③ Enter a number into Box N
     as those to be entered in   ④   N × C → W
     instruction ③.   ⑤   W + A → R
                                          ⑥   OUTPUT R
H   Compare your answers to G   —⑦   GO TO ③
     with those in exercise B.   ⑧   STOP
     What do you notice?

I   Change instructions ① and ② in the program to 3 → A, 1 → C.
     Work through the program as in G. What do you notice?

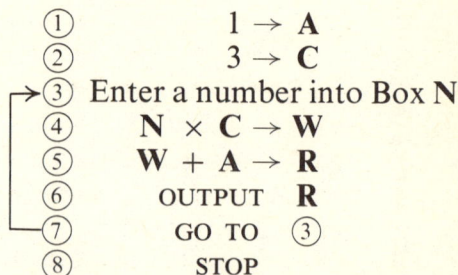

A   Work through the program on the right.
    Make a table with two columns to show
    each output from **N** and **R**.
    Stop when you have filled in ten pairs.
    What do you notice about your results?

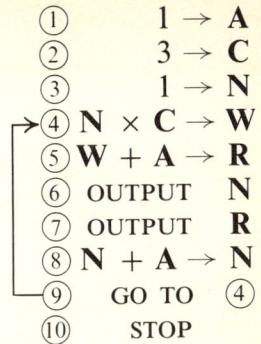

B   If the number in the first column in
    exercise A is *n* what would be the number
    in the second column?

①	1 → **A**
②	3 → **C**
③	1 → **N**
④	**N** × **C** → **W**
⑤	**W** + **A** → **R**
⑥	OUTPUT **N**
⑦	OUTPUT **R**
⑧	**N** + **A** → **N**
⑨	GO TO ④
⑩	STOP

C   Write down a statement showing the relationship between the
    numbers in the **R** column and the numbers in the **N** column.

If you wanted to show a typical member of the set of ordered pairs
in A above, you could use *n* to stand for the number in the
first column. In this case the number in the second column is $3n + 1$.
The typical ordered pair is $(n, 3n + 1)$ where *n* can be any number
from 1 to 10 inclusive.

D   Choose a suitable letter and use it to describe, in a single
    ordered pair, the relationship:

1. (minutes, litres pumped)

2. (number of stamps bought,
   total cost)

3. (litres bought, total cost) for petrol at 12p per litre
4. (time in seconds, length of tape recorded) in making a tape
   recording at a tape speed of 19 centimetres per second
5. (first number, second number) for two numbers whose sum is 8
6. (number of edges, number of diagonals from one vertex) for a set
   of polygons
7. (number of players, time taken) for an orchestra to play a piece
   which takes one player 10 minutes

Transformations

A The yellow shape above is being flipped over about the dotted line.
Its outline before and after are drawn to make a shape.
Make a copy of the yellow shape from cardboard (W/s 36).
Place it on a sheet of paper and draw round its outline in blue.
Flip the shape over about the dotted line. Draw round its outline
again, in green.
What do you notice about the shape made from the blue and
green outlines you have drawn?
Mark any lines of symmetry.

B Repeat A for the shape (W/s 36):

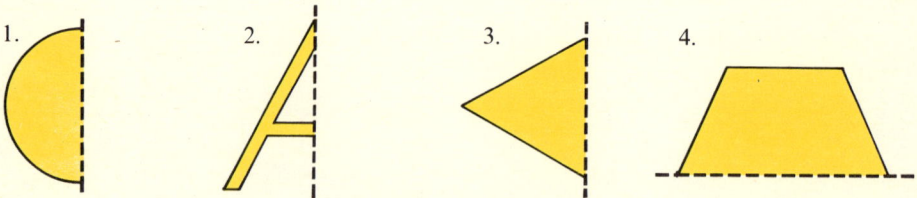

1.   2.   3.   4.

C For each part of B name the shape made by the two outlines.

D Make a regular octagon from cardboard (W/s 36).
Place it on a large sheet of paper. Draw round its outline.
Flip the octagon over about one edge and draw round its outline
again.
Make a pattern by doing this several times using different edges.

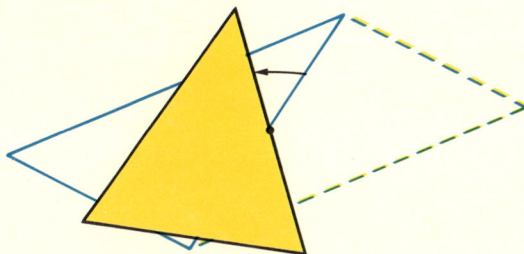

A   The yellow triangle is being given a half turn about the black dot.
     Its position before is shown by the blue outline.
     Its final position is shown by the dotted green outline.
     Make a copy of the yellow triangle from cardboard (W/s 37).
     Place it on a sheet of paper and draw round its outline in blue.
     Rotate the triangle a half turn about the marked point and
     draw round its outline again in green.
     What can you say about the shape made from the blue and green
     outlines you have drawn?

B   Repeat A using the middle point of another edge as the point about
     which to make the half turn.
     Do you finish with the same shape as in A?

C   Repeat A for the shape (W/s 37):

     1.   2.   3.   4.

     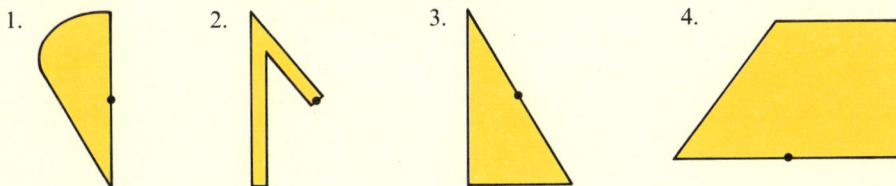

D   For each part of C name the shape made by the two outlines.

E   Make an irregular quadrilateral from cardboard (W/s 37).
     Place it on a large sheet of paper. Draw round its outline.
     Rotate the quadrilateral through a half turn about the middle point
     of each edge. Make a pattern by continuing in this way.

A   Mark a point P near the centre
    of a piece of paper. Near P
    draw a curly shape like the
    blue one shown on the right.
    Lightly pin a sheet of tracing
    paper to P. Trace the shape.
    Rotate the bottom sheet of
    paper a quarter turn anti-
    clockwise. Retrace the shape.

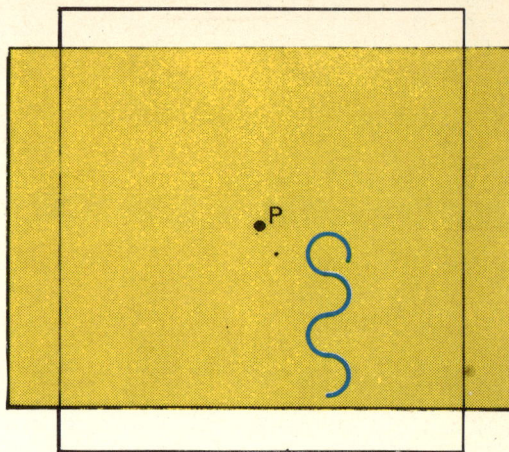

    On the tracing paper you will
    now have the positions of the
    shape before and after the
    quarter-turn rotation. The new
    position of the shape is called
    its **image** after a 90°
    anticlockwise rotation.

B   Look at the pattern on the
    right.
    How do you think it was made?

C   Use your drawing from
    A and rotate the bottom
    sheet of paper another 90°
    anticlockwise from its
    last position.
    Retrace the blue shape.
    Does this help you to
    decide how the pattern on
    the right was drawn?

D   Complete the pattern that you have been drawing in A and C.
    Which part of the pattern is the image of the blue shape after
    an anticlockwise rotation of: 1. 180°?   2. 270°?
    What kind of symmetry has your pattern?

    A pattern which fits onto itself four times when rotated through a
    complete turn has rotational symmetry of order four.

E   Using the ideas on this page make patterns with rotational
    symmetry of order: 1. four   2. two   3. three

The green leaf is the image
of the blue leaf after
a rotation of 180°
about the point P.

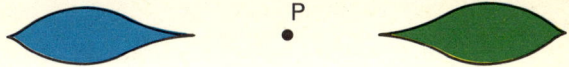

The point P is called the **centre of rotation**.

A   Make an accurate copy (W/s 35)
of the drawing shown on the
right. Where do you think is the
centre of rotation, which makes
the green shape the image of
the blue shape after a rotation
of 180°?
Check by using tracing paper.

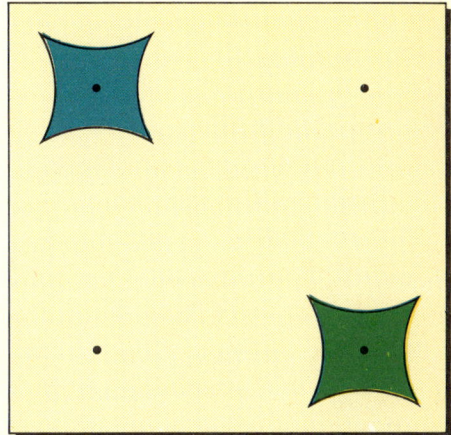

B   Try to find another centre of
rotation, for the shapes in A,
which makes the green shape the
image of the blue shape after
a 90° rotation, anticlockwise.

C   Try to find yet another centre of rotation for the shapes in A
that again makes the green shape the image of the blue shape.
Which rotation operator is needed for this?

D   Make an accurate copy (W/s 35)
of the blue shape drawn on the right.
Mark the point P on your drawing.
Lightly pin a sheet of tracing paper to P.
Trace the blue shape. Mark LM.
Rotate the bottom sheet of paper
60° clockwise.
Retrace the blue shape.
Mark the new position of LM.
Extend LM on each shape
until they meet.
Measure the angle between
the two lines.
Through what angle did
LM turn?

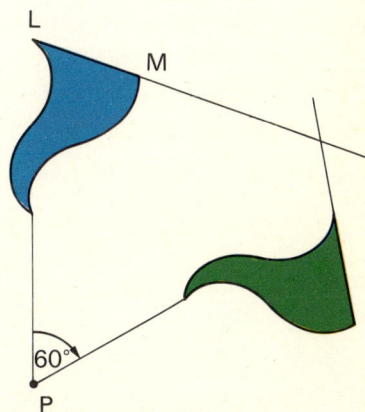

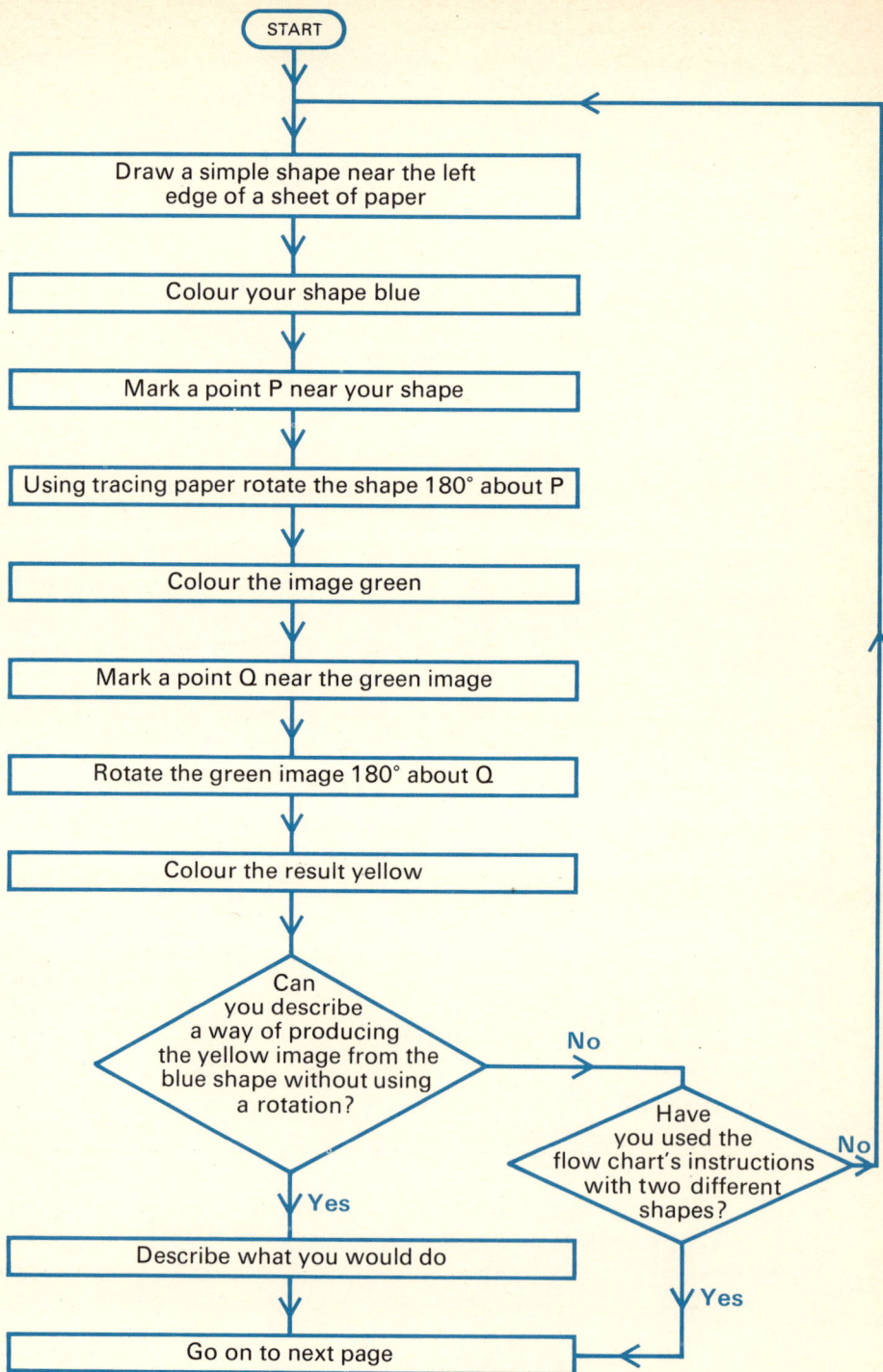

A

START

Draw a simple shape near the left edge of a sheet of paper

Colour your shape blue

Mark a point P near your shape

Using tracing paper rotate the shape 180° about P

Colour the image green

Mark a point Q near the green image

Rotate the green image 180° about Q

Colour the result yellow

Can you describe a way of producing the yellow image from the blue shape without using a rotation?

No

Have you used the flow chart's instructions with two different shapes?

No

Yes

Describe what you would do

Yes

Go on to next page

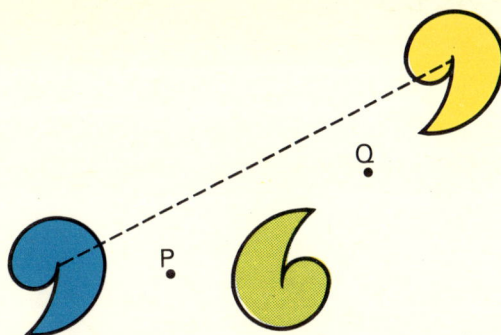

The result of a drawing made after following the flow chart on the last page is shown above.

The blue shape was first rotated 180° about P.

The image was then rotated 180° about Q to produce the yellow shape.

The yellow shape could also be obtained by simply moving the blue shape along a straight line.

A move like this along a straight line is called a **translation**.

A   Measure the distance between the tail of the blue shape and the tail of the yellow shape. Repeat for other points on the two shapes. Measure the distance between P and Q.

What do you notice?

B   Repeat A for some of the drawings which you made on page 70.

C   The drawing of the pattern above was started by rotating the first blue swan 180° about P.

How could the rest of the pattern be made using translations?

D   Use a rotation followed by several translations to make your own pattern like the one above.

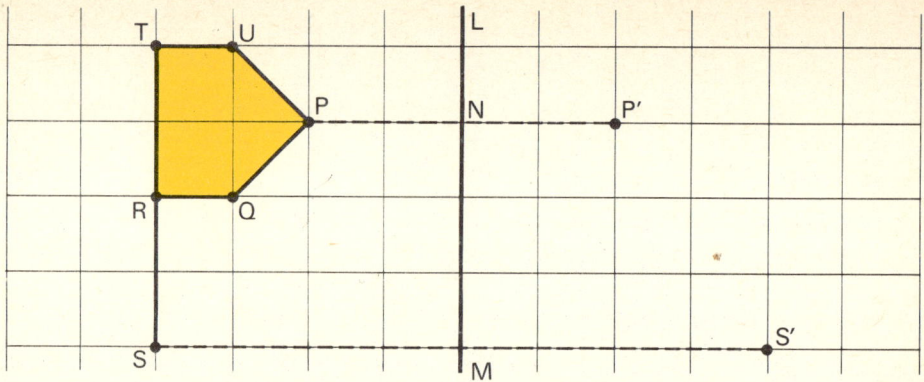

A   Copy the shape SRTUPQ above (W/s 38) and mark the line LM on your diagram. Draw a line from P at right angles to LM. Extend this line to P′, the same distance as PN on the opposite side of LM.
Using this rule, mark S′, the image of S.
Repeat for Q, R, T, and U to get their images Q′, R′, T′, and U′.
Join in turn S′, R′, T′, U′, P′, Q′. What do you notice about this shape?
Describe how to move the yellow shape to its image S′R′T′U′P′Q′.

B

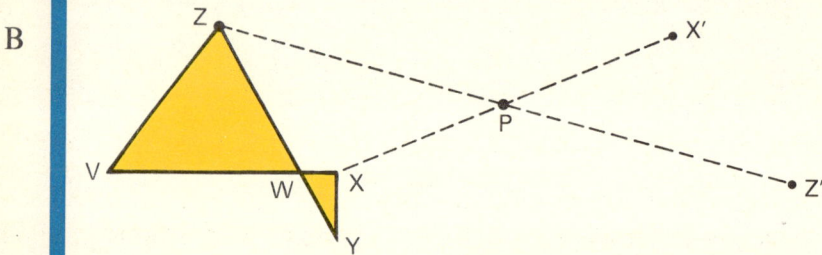

Copy the shape VWXYZ (W/s 38) and mark a point P as shown.
Join XP and extend this line to X′, so that PX′ = XP.
Using this rule mark Z′, the image of Z.
Repeat for the points V, W, and Y to get their images V′, W′, and Y′.
Join in turn V′, W′, X′, Y′, Z′. What do you notice about this shape?
Describe how to move the yellow shape onto its image.

C   Copy the shape LMNK (W/s 38) and mark P.
Join PL and extend to L′, so that PL′ = 2PL.
Using this rule, mark the images of K, M, and N.
What can you say about L′M′N′K′?

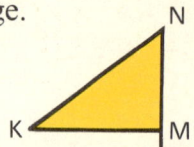

D   If PX′ = 2XP in exercise B, what is the new image (W/s 39)?

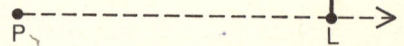

A   Draw a simple shape on the left-hand side of a sheet of paper.
     Using two 180° rotations, translate the shape 5 cm across the
     paper.

B   Make a copy (W/s 39) of the
     drawing shown on the right.
     Translate the yellow shape 8 cm
     in the direction of the line RS.
     Colour the image green.
     Rotate the green image 180°
     about S.
     Colour this new image blue.
     If the blue image is rotated
     180° back to the original shape
     where would the centre of rotation
     have to be?

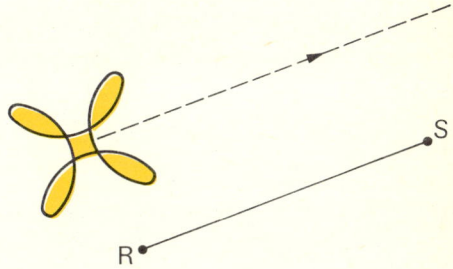

An operator which alters either the position, or the size of a
shape is called a **transformation** operator.

Rotation, translation, and enlargement are three types of
transformation.

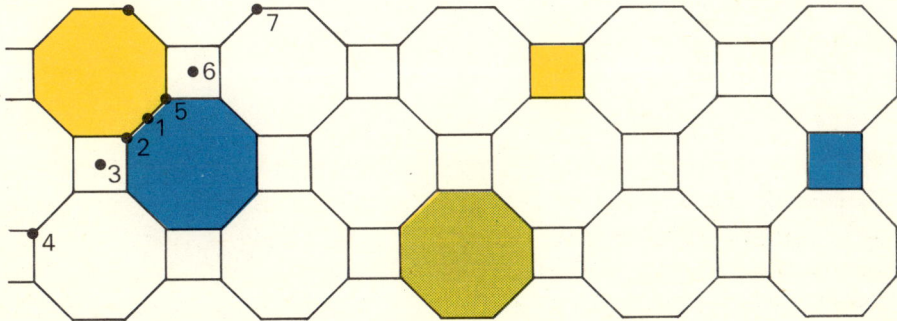

C   Look at the tessellation of squares and octagons.
     In what ways can you move the yellow square so that it has the
     blue square as its image? Describe each way carefully.

D   Each of the seven dots marked on the tessellation above is a
     possible centre of rotation for moving the yellow octagon onto the
     blue octagon. What is the angle of rotation in each case?

E   The green octagon is the image of the blue octagon. Investigate.

Sets and letters

A   Look carefully at the set of sports equipment above.
     Choose a subset and write down a list of its members.
     Describe the subset you have chosen.

B   Repeat A for a second subset.

     To save time you can use a letter to stand for each of your subsets.
     You could use C for {cricket equipment in the picture},
               and P for {protective clothing in the picture}.

C   Write down a list of the items which belong to both C *and* P.
     This set is the intersection of C and P.
     Copy and complete the statement C ∩ P = {   }.

D   If H = {items in the picture used for hitting a ball}
        I = {items in the picture used for indoor sport}
        W = {items in the picture made from wood}
     list the members of: 1. H   2. I   3. W

E   Copy and complete the statement: 1. H ∩ I = {   }
     2. I ∩ W = {   }   3. H ∩ I ∩ W = {   }

A   Using the set of sports equipment on the opposite page, write down a list of those items in the picture which fit the description 'made from wood' *but not* 'used for hitting a ball'.

The symbol used for *but not* is \.
The set in exercise A could be described as W \ H.

B   Copy and complete the statement: 1. I \ H = {   }.
                                        2. W \ C = {   }.

Whole numbers from 1 to 25

2   3   5   6   7   10   11

1   4   8   24   13

9   16   12   14   15

25   20   17

23

22   21   19   18

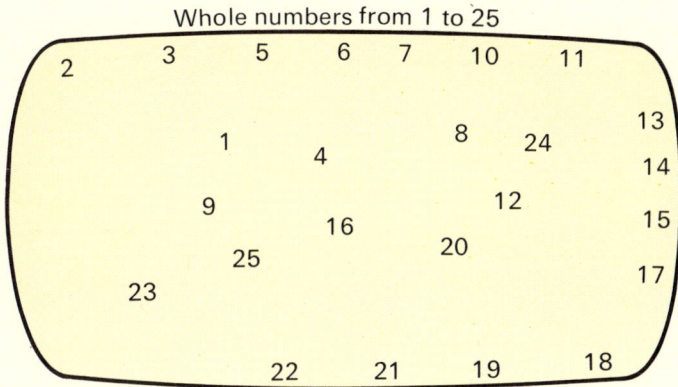

C   Copy the diagram above (W/s 40).
    S = {square numbers}. Draw a blue loop round this set.
    M = {multiples of 4}.   Draw a yellow loop round this set.

D   Write down the numbers which belong to S *but not* M.
    Copy and complete the statement, S \ M = {   }.
    How many numbers are in this set?

E   Copy and complete the statement, M \ S = {   }.
    How many numbers are in this set?

F   Does S \ M = M \ S?

G   Write down the numbers which belong to both S *and* M.
    Copy and complete the statement, S ∩ M = {   }.

H   Copy and complete the statement, M ∩ S = {   }.
    Does S ∩ M = M ∩ S?

I   Describe the set of numbers in the diagram which are not in M.
    What is the intersection of this set with the numbers in S?

Often it is necessary to state the set of items being considered.
If you were asked to write down the set of whole numbers greater
than 8, you might give {9}, or {9, 10, 11; 12}. In the first case
you would be correct if you were talking about single digit numbers.
In the second case you would be correct if you were talking about
the numbers on a clock face.
The set you are considering is usually referred to as the **universal** set.
In the diagram below the universal set is {letters of the alphabet}.

Letters of the alphabet

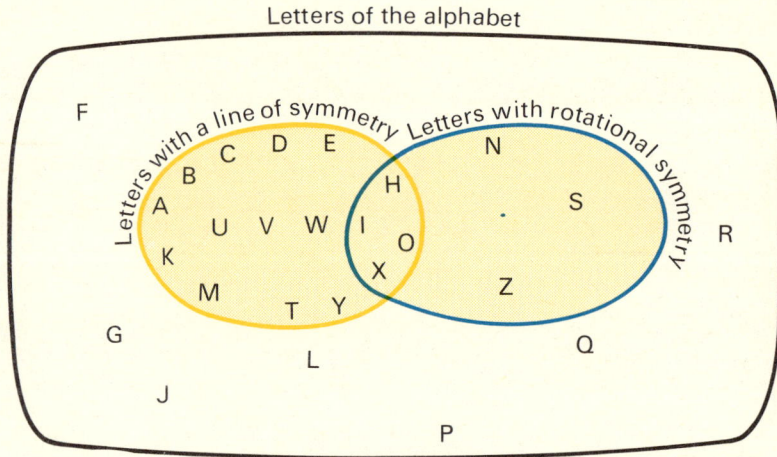

A   If L = {letters with a line of symmetry}
      R = {letters with rotational symmetry},
    list the members of: 1. L   2. R.

B   Write down the members of: 1. L ∩ R   2. L \ R   3. R \ L.

C   Write the members of the three sets in B as one set.
    What can you say about the members of this new set?

    The set you made in C consists of all those letters which have
    either a line of symmetry *or* rotational symmetry, or both.
    This set is called the **union** of the two sets.
    On the diagram above it is all those numbers inside the
    regions shaded yellow. It is written as L ∪ R.

D   If your universal set is {whole numbers from 1 to 25}, write down
    those numbers which are *either* multiples of 4 *or* square numbers.
    What is the union of: 1. {multiples of 4} and {square numbers}?
                              2. {factors of 12} and {prime numbers}?

A   Write down a list of all the people in the room you are in who are *either* wearing glasses *or* who have a tie on.

B   If G = {people in the room wearing glasses}
       T = {people in the room with a tie on},
how many people are in:
1. G?   2. T?   3. the union of G and T, that is $(G \cup T)$?

C   Is your answer to B3 a smaller number than the sum of your answers to B1 and B2? Explain why this is so.

D   Draw a Venn diagram representing the people in the room, with regions representing G and T. Label your diagram carefully to show how many people there are in each of the four regions. How many people are in $G \cap T$?

E   Write down the members of: 1. $G \setminus T$ 2. $G \cap T$ 3. $T \setminus G$.

F   Compare your answers to E with your answer to A.
What do you notice?

G   If P = {numbers which contain the numeral 3}
       Q = {numbers which contain the numeral 5}
       R = {numbers which contain two numerals the same}
       S = {numbers whose separate numerals add up to 6}
       T = {numbers exactly divisible by 11}
write down, for numbers from 11 to 36, the members of:
1. P   2. Q   3. R   4. S   5. T

H   Write down the members of:
1. $P \cap Q$ 2. $P \cap R$ 3. $P \cap S$ 4. $Q \cap S$
5. $Q \cap R$ 6. $R \cap T$ 7. $P \cap R \cap S$ 8. $P \cap Q \cap S$.

I   Write down the members of: 1. $Q \setminus S$ 2. $R \setminus P$ 3. $R \setminus S$.

J   Write down the members of:
1. $Q \cup S$ 2. $R \cup S$ 3. $P \cup R$ 4. $Q \cup S \cup R$ 5. $P \cup Q \cup R$.

K   Can you find $S \setminus R \setminus Q$?

L   Compare your answer to exercise K with your friends. Do you agree?
What is $(S \setminus R) \setminus Q$? What about $S \setminus (R \setminus Q)$?
Does the same problem occur with $8 \div 4 \div 2$?

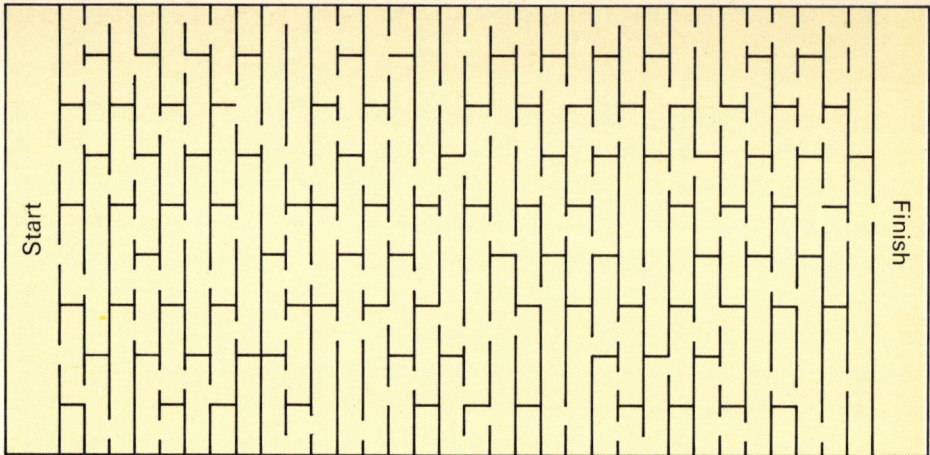

A   Look at the maze and try to find a way from start to finish.
Can you find a solution to this problem?
Is there more than one solution?

You probably found a number of solutions to the problem above.
This set of solutions is called the **solution set** for the problem.

B   Here is another problem $3 + \square = 8$.
Can you find a solution to this problem?

In B the *solution set* is the set of numbers that can be put in
the $\square$ to make the statement true. The only member of the set is 5.

C   If you are allowed to use any number from $\{1, 2, 3, 4, 5, 6, 7, 8, 9\}$,
find the solution set for: 1. $\square > 6$ 2. $\square < 3$ 3. $5 + \square = 9$
4. $3 + \square > 8$ 5. $6 + \square < 8$.

D   For each part of C write down the number of members in the
solution set. Which set has the most members?

E   If you are allowed to use any number from $\{1, 2, 3, 4, 5, 6, 7, 8, 9\}$
find the solution set for: 1. $\square > 4$ 2. $\square < 7$.
What is the intersection of these two sets?

The numbers in the intersection are those numbers which make
the statements $\square > 4$ *and* $\square < 7$ true at the same time.
In this case only 5 and 6 belong to both solution sets.

Use {whole numbers from 1 to 20} as the universal set for the answers to each exercise on this page.

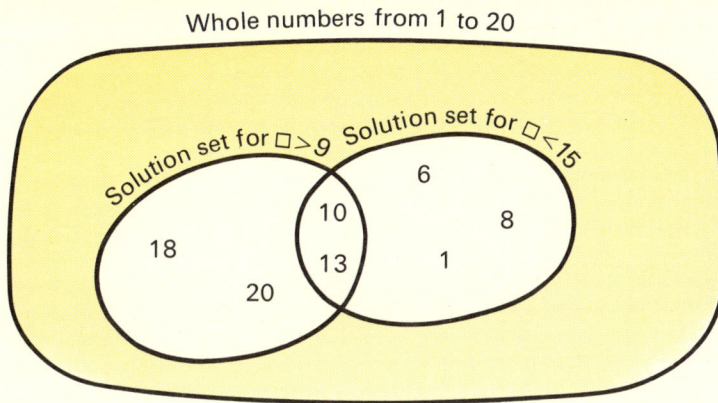

Whole numbers from 1 to 20

Solution set for □ > 9   Solution set for □ < 15

18   20   10   13   1   6   8

A   Copy and complete the diagram above (W/s 40).

B   What numbers can be put in the □ to make a true statement for:
1. □ > 9?   2. □ < 15?

C   For what numbers are the statements □ > 9 *and* □ < 15 true at the same time?

D   For what numbers is the statement □ > 9 true *but not* □ < 15?

E   Which numbers belong to the solution set for □ < 15 *but not* to the solution set for □ > 9?

F   Find the solution set for: 1. □ × 4 < 23   2. 20 − □ < 17

G   For what numbers are the statements □ × 4 < 23 *and* 20 − □ < 17 true at the same time?

H   Which numbers belong to the solution set for □ × 4 < 23 *but not* to the solution set for 20 − □ < 17?

I   For what numbers is the statement 20 − □ < 17 true *but not* the statement □ × 4 < 23?

J   For what numbers is *either* □ > 13 *or* □ < 4?

K   Find the union of the solution sets for 2 + □ < 7 and □ × 3 > 50.

Probability

The children at Manton School are excited. The school's new swimming pool is to be officially opened next week. A child is to dive in and swim two lengths as part of the opening ceremony. The names of the eight best swimmers in the school are to be placed in a hat and the person whose name is drawn out will swim the two lengths. The names of the eight swimmers are shown above.

A   What are the chances that Rachel Williams' name will be drawn from the hat?

Eight names are put into the hat. Only one name will be drawn out. Rachel's chance of selection is one in eight.
The **probability** of her selection can be written as **1 in 8**.

B   What is the probability that Alan Fairfax will be selected?

C   Is it more likely that a girl rather than a boy will be selected? Why?

The probability of a girl being selected to swim in the opening ceremony for the swimming pool can be shown as the fraction:

$$\frac{\text{number of girls in the set}}{\text{number of children in the set}} \qquad \text{The probability is } \frac{5}{8}$$

D   Write, as a fraction, the probability of a boy being selected for the swim.

A   You need a white cube.
     Colour three faces blue
     and two faces yellow.

B   If you roll the cube
     what is the probability
     of a blue face coming
     on top? Write your
     answer as a fraction.

C   Write, as a fraction, the probability of the cube landing with:
     1. a yellow face on top.   2. a white face on top.

D   If you roll the cube 72 times, how many times do you think
     the face on top will be: 1. blue?   2. yellow?   3. white?

E   Try the experiment suggested in D.
     Did you get the result that you forecasted?

     When you make a forecast, based on probability ideas, you are
     saying what is *likely* to happen. You are not saying what will
     happen.

F   Look at the set of cards on
     the right. They are mixed up
     and placed face down on the
     table.
     One card is to be picked out.
     What is the probability that
     the chosen card will be:
     1. a black card?
     2. a diamond?   3. a spade?
     4. a card with an even number of pips?
     5. a black card with an odd number
        of pips?

G   In a pack of 52 playing cards, one card is picked out. What is
     the probability that the chosen card will be:
     1. an ace?   2. a club?   3. a picture card?   4. a red 10?

A   Find four yoghurt cartons and
a marble.
Label the cartons P, Q, R, S.

B   Ask a friend to put the marble
under one of the cartons, whilst
you look the other way.
Now you guess which carton the
marble is under. What is the
probability of your being right?

C   Repeat the activity in **B**.
Were you right both times?
What, do you think, is the
probability of this happening?

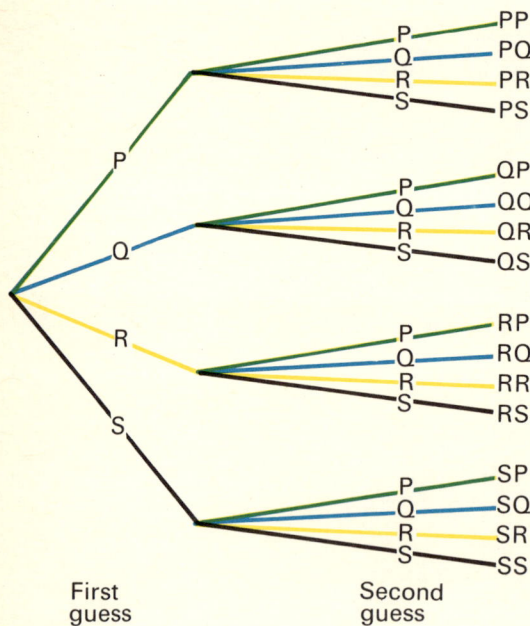

The first time the marble could
have been under P, Q, R, or S.
The probability of guessing
correctly is 1 in 4.
Similarly, the second time, the
marble could have been under
P, Q, R, or S. Once again the
probability of guessing correctly
is 1 in 4.
The various pairs of possible
results can be shown in a
tree diagram, as shown on the
left.
Only one branch end can show
the actual result, and so the
probability of guessing
correctly both times is 1 in 16.

First guess   Second guess

D   What would have been the probability of guessing correctly in
B and **C** if there had been: 1. 5 cartons?   2. 6 cartons?

E   When tossing a 2p coin and a 10p coin, what is the probability of
getting two heads?

A Work with your friend.
Each of you have 32 turns
of the activities in B and C
at the top of page 84. Record
your results in a table, like
the one shown on the right.

Turn number	1	2	3	4	5
First guess	✕	✕	✓	✓	✕
Second guess	✕	✕	✕	✓	✕

B How many times were you right for both guesses in a turn, in A?
What about your friend? Are the results what you expected?

C You need a box with an equal number of blue and yellow counters in
it. Pick a counter without looking and record its colour. *Put it back
in the box*. Pick a second counter and record its colour.
Do you think the two selected counters are more likely to be:
1. the same, or different colours?
2. both blue, or one of each colour?
3. both yellow, or one of each colour?

D Repeat the activity in C 60 times.
Record your results in a table.
Are your results what you expected?

E The tree on the right shows the
possible ways in which the counters
could have been chosen in D.
What is the probability of picking:
1. two blues?   2. two yellows?
3. one of each colour?

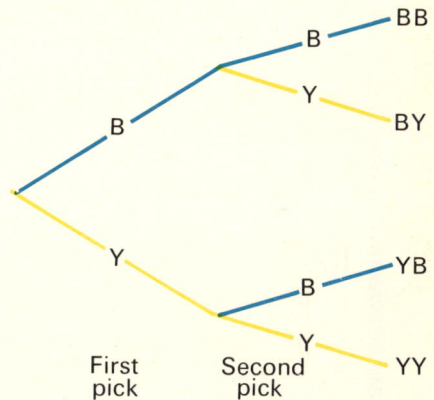

First pick   Second pick

F If you carry out a similar experiment
to the one in C, but this time picking
three counters, what would be the probability of picking:
1. three blues?   2. two blues and a yellow?
3. two yellows and a blue?   4. three blues?
Would drawing a tree help you?

G If you tossed three coins, what is the probability of having:
1. 3 heads?   2. 2 heads and a tail?   3. 2 tails and a head?

H How would you expect your answers to C to be different for a
box with twice as many blue counters as yellow counters?

Twenty bingo counters

Multiples of 3

Multiples of 11

31
43
80
30
33
11
40
66
44
22
60
99
77
55
59
90
88
17
20
23

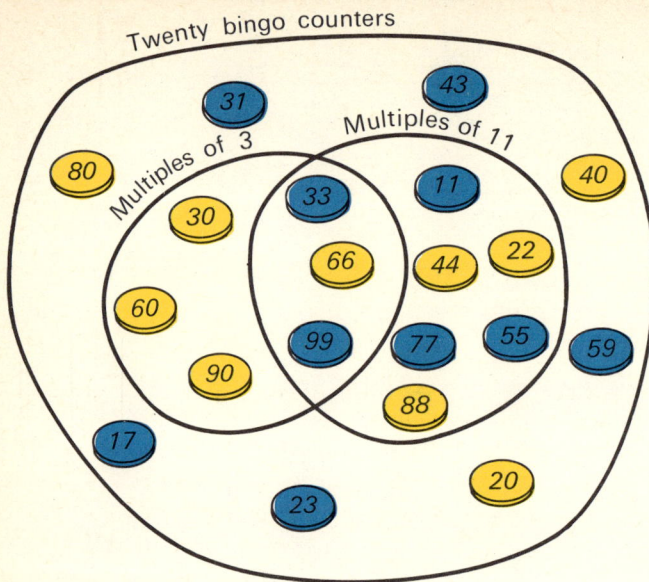

A   Look at the Venn diagram above. In a game of bingo these were
     the first twenty counters picked out of the bag.
     These twenty counters are now placed in a second bag.
     What would be the probability of picking from the second bag:
     1. a multiple of three?   2. a multiple of eleven?
     3. a multiple of three *and* eleven?
     4. a multiple of three *but not* eleven?
     5. a multiple of eleven *but not* three?
     6. a counter which is neither a multiple of three, nor one of eleven?
     7. a prime number?
     8. a multiple of fourteen?

B   Re-sort the twenty counters in A and show in a Venn diagram:
     1. yellow counters *and* multiples of five
     2. multiples of four *and* multiples of ten

C   Use your Venn diagram in B to decide what the probability is
     of picking from a bag:
     1. a yellow counter *and* a multiple of five
     2. a yellow counter *but not* a multiple of five
     3. a multiple of four *or* a multiple of ten

D   Answer the questions in A for:
     1. a full set of bingo counters (i.e. 1–99)
     2. a pack of 52 playing cards

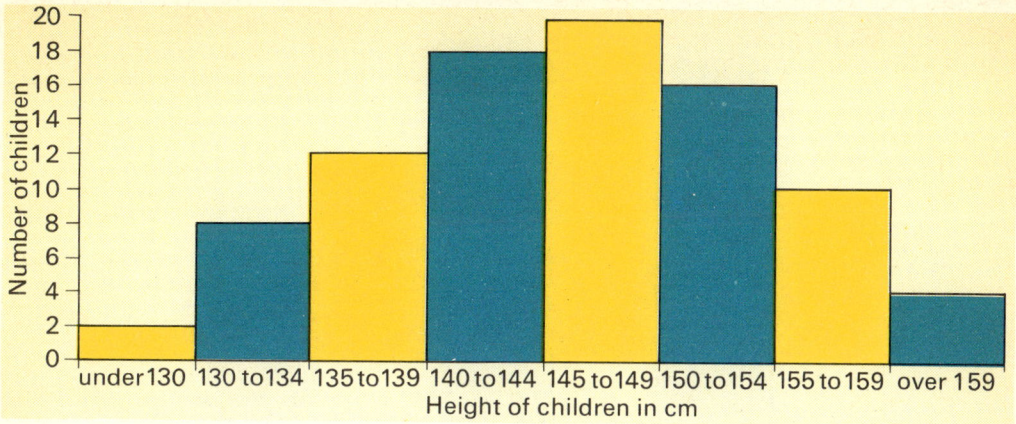

A survey was carried out amongst the second year of a small comprehensive school. The results are shown in the graph above.

A
1. How many children were involved in the survey?
2. Which was the most common set of heights?

B
Show the number of children in the most-common set as:
1. a fraction   2. a percentage,   of the number of children in the survey

C
If one member of the second year wins the year prize, what is the probability that his height is:
1. 140 cm to 144 cm?   2. 155 cm to 159 cm?   3. under 130 cm?

The pie chart shown below indicates the result of a survey, in which 1200 people, who watched BBC1 television on Friday 23rd November 1973, were asked which of five programmes they preferred.

D
How many of the 1200 viewers preferred each programme?

E
If you had picked out any one of the 1200 viewers, what is the probability that he preferred:
1. Miss World?   2. News at 9 p.m.?

F
Look in the national papers.
Find out all you can about the various national opinion polls.
How are the people chosen for the surveys?

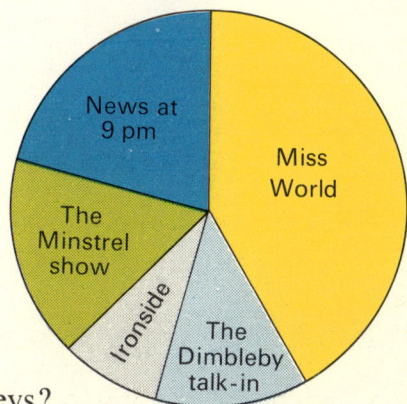

Using graphs

A graph to show how long some animals have lived

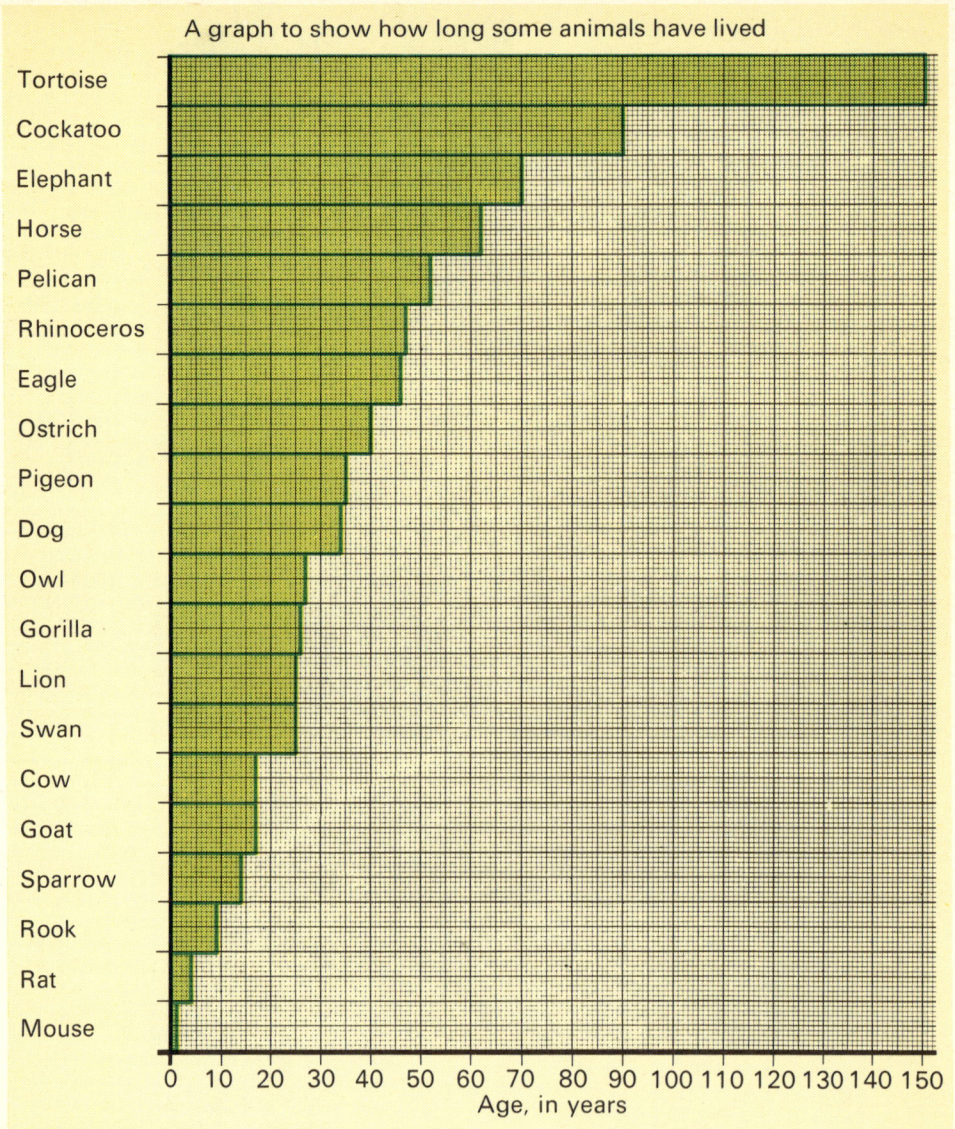

Animals listed top to bottom: Tortoise, Cockatoo, Elephant, Horse, Pelican, Rhinoceros, Eagle, Ostrich, Pigeon, Dog, Owl, Gorilla, Lion, Swan, Cow, Goat, Sparrow, Rook, Rat, Mouse

Age, in years: 0 10 20 30 40 50 60 70 80 90 100 110 120 130 140 150

A   Look at the graph. Write down anything you find interesting in it.

B   Do you think that the age to which an animal lives depends upon:
     1. its size?   2. whether it is wild or domesticated?

C   Find out all you can about the ages to which other animals have
     lived. Show your results on a graph.

A graph showing marks out of 30 as percentages

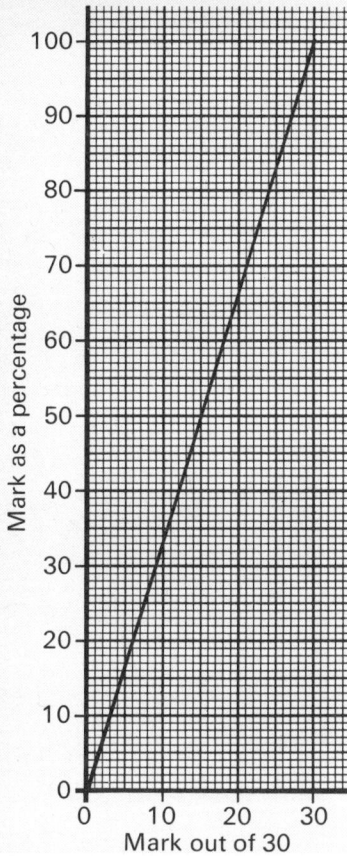

A graph showing marks out of 55 as percentages

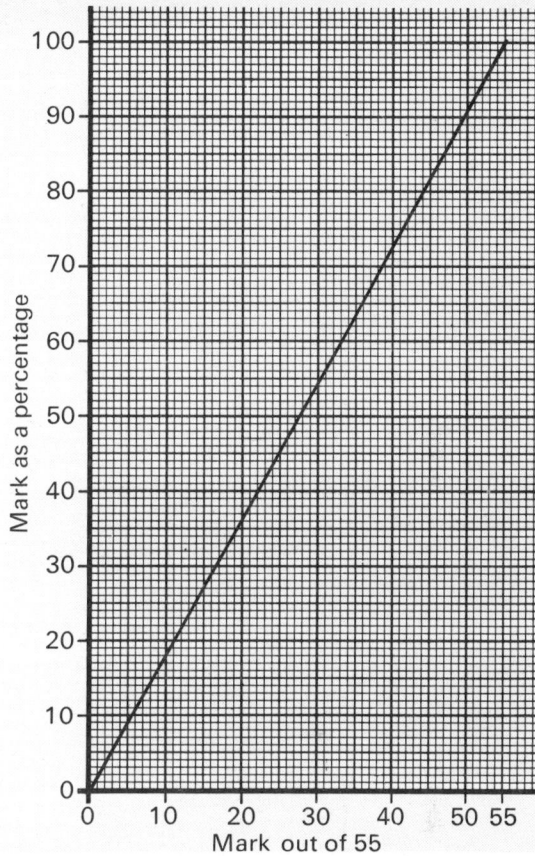

A   Look at the graph, on the left above.
    Check that it is correct by looking at the percentages for marks
    such as: 30 out of 30 (100%), 15 out of 30 (50%), 0 out of 30.
    In the same way look at and check the graph on the right.

B   Use the graph on the left to change to a percentage a mark, out
    of 30, of: 1. 24   2. 9   3. 17   4. 27   5. 13   6. 7   7. 29

C   Use the graph on the right to change to a percentage a mark, out
    of 55, of: 1. 20   2. 35   3. 42   4. 8   5. 50   6. 16   7. 46

D   If you have a calculating machine check your answers to B and C.

A graph to show the length of the longest day at various latitudes

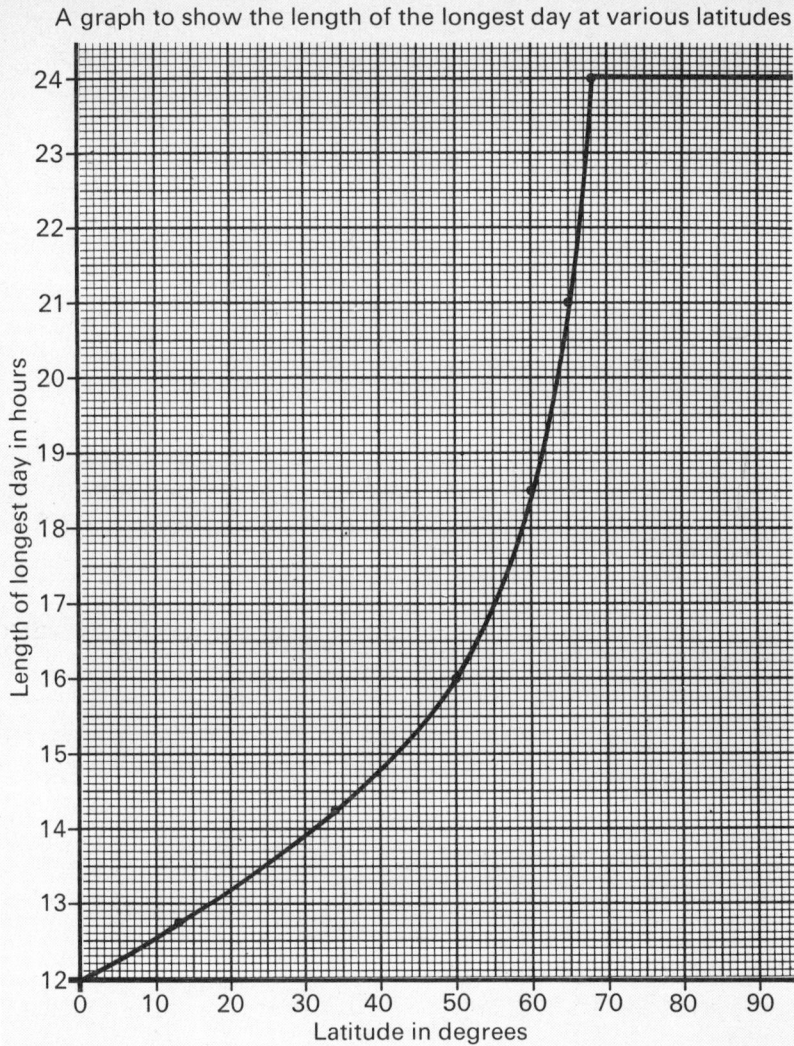

Length of longest day in hours / Latitude in degrees

A   Look at the graph above. Can you explain why there is a straight piece on the graph?

B   Use the graph and your atlas to find the length of the longest day at: 1. London   2. Bombay   3. New York   4. Murmansk

C   Find a place where the length of the longest day is about:
     1. 17 hours   2. 14 hours   3. 23 hours   4. 20 hours

A graph to show the fees for Postal Orders

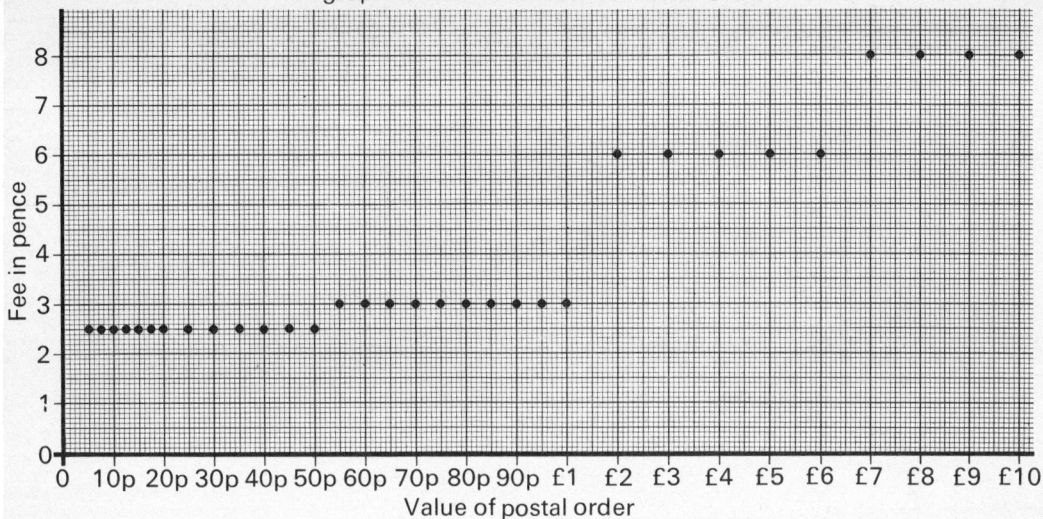

A   The above graph shows the fees charged (in 1973) for all the various postal orders sold by the Post Office.
Write down all the information you can get from it.

B   What do you notice about the way in which the axis across the bottom of the graph is labelled? Why do you think it is done in this way?

C   Do you think it would be correct to draw lines through the sets of points? Give reasons for your answer.

D   Use the graph to find the fee for a postal order of value:
1. 5p   2. 7½p   3. 15p   4. 35p   5. 85p   6. £5   7. £9

E   The value of a postal order may be increased by affixing not more than two stamps up to a value of 4½p. Use this information and the graph to find the least total cost of sending by post:
1. £1·58   2. £3·54   3. £12·88   4. £2·14   5. £21·82.

Equations and graphs

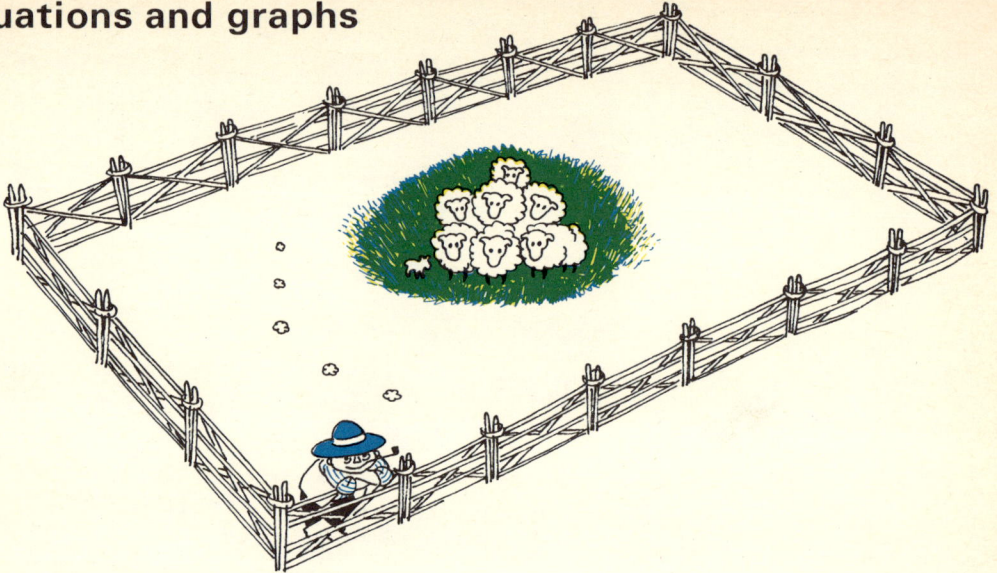

A Farmer McTavish has twenty fence panels to make a rectangular sheep pen. One way he could do this is shown above.
How many fence panels are used for the long edge of the pen?

B Rearrange the twenty fence panels to make another rectangular sheep pen. How many panels did you have in the long edge?

C In how many different ways can you arrange twenty fence panels to make a rectangular sheep pen?
In each case write down the number of panels used for each edge.

D Write down a statement about the number of panels used altogether for a long edge and a short edge.

You probably found that for each pen the number of panels used for a long edge plus a short edge was ten. Another way of writing this is by using a $\square$ and $\triangle$ statement: $\square + \triangle = 10$
If 7 is put in the $\square$, 3 must be put in the $\triangle$ to make the statement true. If 3 is put in the $\square$, then 7 must be put in the $\triangle$.

E Write down as many pairs of whole numbers as you can that make the statement true:
1. $\square + \triangle = 8$ 2. $\square + \triangle = 3$ 3. $\square \times \triangle = 12$

F Which of the pens in C would hold the greatest number of sheep?

A graph to show the lengths of the edges of a rectangle made from 36 squares

Length in cm of side edge of rectangle

Length in cm of bottom edge of rectangle

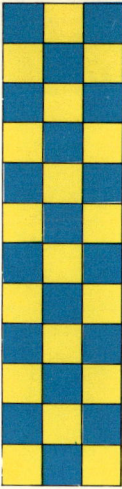

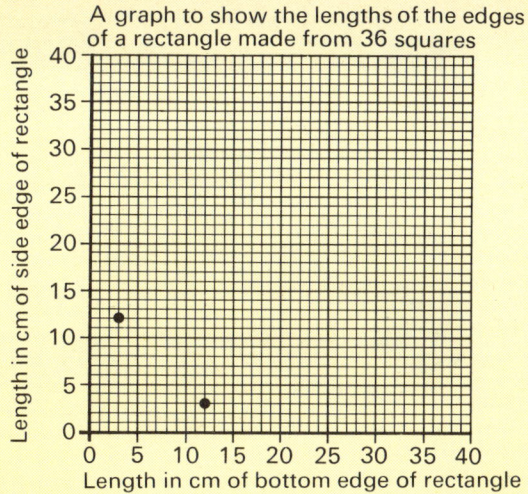

A   Use 36 one-cm squares (W/s 41). Arrange them in the shape of a rectangle. Write down the length of each edge, in a table like the one below (W/s 41). Write down the area of your rectangle.

Length in cm of bottom edge	3					
Length in cm of side edge	12					
Area in cm² of rectangle	36					

B   Rearrange your 36 squares to make another rectangle. Record the length of each edge and also the area of the rectangle in your table.

C   How many different rectangles can you make with 36 squares? For each, record the information in your table like the one above. What do you notice about your results?

D   Write down a ☐ and △ statement about the lengths of the edges of the rectangles you have made in C.

E   Record as a set of ordered pairs, the pairs of numbers that make your ☐ and △ statement in D true. One of them is (3, 12).

F   Copy the graph at the top of the page (W/s 41) and show on this graph the set of ordered pairs you found in E.

G   For each part of E on page 92, draw a graph (W/s 42) and show the ordered pairs you found. What can you say about your graphs?

(1,5)

(3,3)   (4,2)

(2,4)

(5,1)

Look at the diagram above. If one die is blue and the other is yellow, the ordered pairs could be (score on the blue die, score on the yellow die). A □, △ statement for this is □ + △ = 6. The score on the blue die would be the number put in the □. The △ would hold the yellow score.

A   In how many different ways can you make a score of 5, when rolling two dice? Record your answers as a set of ordered pairs.

B   Write down a □, △ statement for your set of ordered pairs in A.

C   Using a blue die and a yellow die, find the ways in which the score on the blue die can be three more than the score on the yellow die. Write these as a set of ordered pairs.
Would the □, △ statement for these ordered pairs be:
□ − △ = 3,   △ − □ = 3   or   □ + △ = 3?

D   Two dice are rolled. The scores have been recorded below. Describe the results in your own words.

1.
| Score on blue die | □ | 1 | 2 | 3 | 4 | 5 | 6 |
| Score on yellow die | △ | 6 | 5 | 4 | 3 | 2 | 1 |

2.
| Score on blue die | □ | 1 | 2 | 3 | 4 | 5 | 6 |
| Score on yellow die | △ | 1 | 2 | 3 | 4 | 5 | 6 |

3.
| Score on blue die | □ | 1 | 2 | 3 |
| Score on yellow die | △ | 2 | 4 | 6 |

E   For each part of D, write a □, △ statement.

As you found on page 76, the use of letters can shorten the description of a situation. If B stands for the score on the *blue* die and Y stands for the score on the *yellow* die then

$$B + Y = 7$$

could be used to describe D1 on the opposite page (i.e. a total score of 7).

A Copy and complete the table on the right (W/s 43) for $B + Y = 8$.

B	6	5		3
Y			4	6

B Copy and complete the table on the right (W/s 43) for $B = Y \times 2$.

B	2		6
Y		2	

C Copy and complete the table on the right (W/s 43) for $B - Y = 1$.

B	6	5		3
Y			3	1

D Write down a B, Y statement for the table on the right.

B	1	3	4	2
Y	4	2	1	3

E Look at the graph on the right. Write down the set of ordered pairs shown. Describe this set using a B, Y statement.

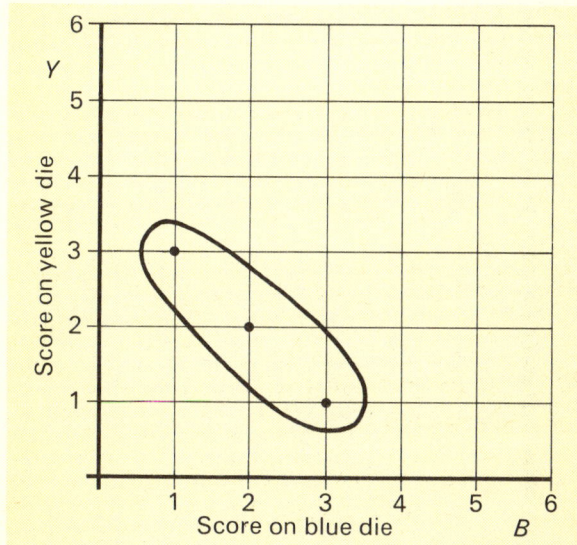

F Copy the graph above (W/s 43) and mark on it, in yellow, the set of ordered pairs for the statement $B = Y$.

G Find the intersection of the sets in E and F.

H What is the probability of scoring 4 when throwing two dice?

A   Record some members of the
set of ordered pairs
(number bought, total cost)
for the batteries shown on the right.

B   Write a statement for A
which begins:
Total cost of the batteries =

C   If you bought n batteries,
what would be the total cost?

D   If n stands for the number of batteries bought, and t stands for the
cost of them in pence, write a statement showing the relationship
between n and t.

E   If you bought m metres of wire what would be the cost?

F   If m stands for the number of metres of wire bought and c pence is
the cost of this wire, write a statement showing the relationship
between m and c.

In F the statement showing the relationship between c and m, is
$c = m \times 4$. This is often called an **equation**.
If m is 6, to make a true statement c must be 24.

G   What number must be written for c to make a true statement if m is:
1. 3?   2. 5?   3. 12?

H   What number must be written for m to make a true statement if c is:
1. 12?   2. 36?   3. 144?

I   If $a \xrightarrow{+3} b$ find the value of b when a is 7.
Write down the equation showing the relationship between a and b.

J   If $p \xrightarrow{\times 4} q$ find the value of q when p is 9.
Write down the equation showing the relationship between p and q.

K   If $x \xrightarrow{\times 4} \cdot \xrightarrow{+3} y$ find the value of y when x is 5.
Write down the equation showing the relationship between x and y.

The relationship between x and y in the arrow graph below:

$$\overset{\times 3}{\xrightarrow{\hspace{2cm}}} \quad \overset{-2}{\xrightarrow{\hspace{2cm}}}$$
$$x \hspace{6cm} y$$

can also be shown as the equation $y = (x \times 3) - 2$.

A Find the value of y in the arrow graph above when x is 4. Check that the ordered pair (x, y) will make the equation a true statement.

B Find two more ordered pairs that will make $y = (x \times 3) - 2$ true.

x and y are members of: {whole numbers from 1 to 36}

The graph shows the set of ordered pairs (x, y) which makes the equation $y = (x \times 3) - 2$ true.

The points in the graph represent {(1, 1) (2, 4) (3, 7) (4, 10) (5, 13) (6, 16) (7, 19) (8, 22) (9, 25) (10, 28) (11, 31) (12, 34)}

This set is called the solution set of the equation $y = (x \times 3) - 2$.

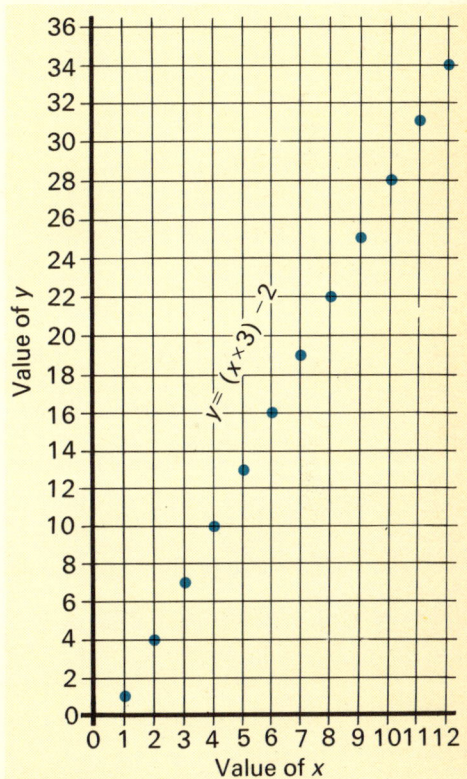

C If x and y are members of {whole numbers from 1 to 12}, find the ordered pairs that belong to the solution set for $y = 12 - x$.
Show this set as a graph (W/s 44).
What difference would it make to your graph if x and y were members of {all numbers from 1 to 12}?

D Repeat C (W/s 44, 45) for:
1. $y = x \times 2$ 2. $y = x - 4$ 3. $y = 24 \div x$

E Repeat C (W/s 46) for: 1. $y = (x \times 2) - 3$ 2. $y = (x \times 3) + 1$

Transformations and co-ordinates

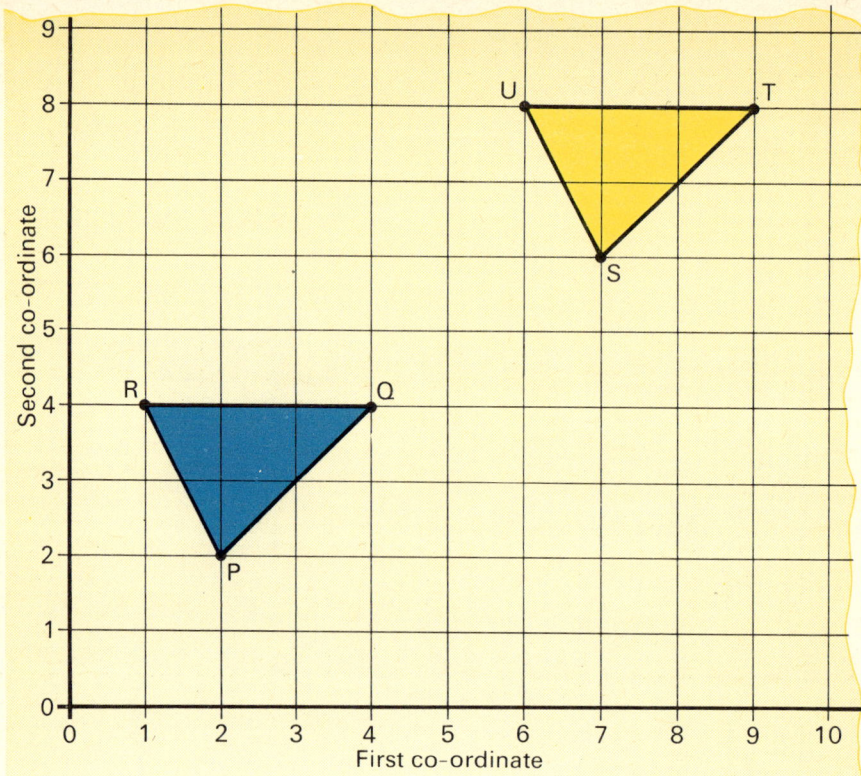

A   Look at the diagram above. The green triangle is the image of
the blue triangle after a translation. Write down the co-ordinates of:
1. P and its image   2. Q and its image   3. R and its image
In what way are the co-ordinates of each point and the co-ordinates
of its image connected?

B   The co-ordinates of the four vertices of a square are
(2, 2), (4, 3), (3, 5), and (1, 4). The square is translated so that
the image of the point (2, 2) is the point (7, 2).
Write down the co-ordinates of each of the other vertices of the
image square. (You may need to draw it on graph paper.)

C   The blue triangle above is flipped over the line RQ.
Write down the co-ordinates of the image of P.

D   The blue triangle above is flipped over the line PQ.
Write down the co-ordinates of the image of R.

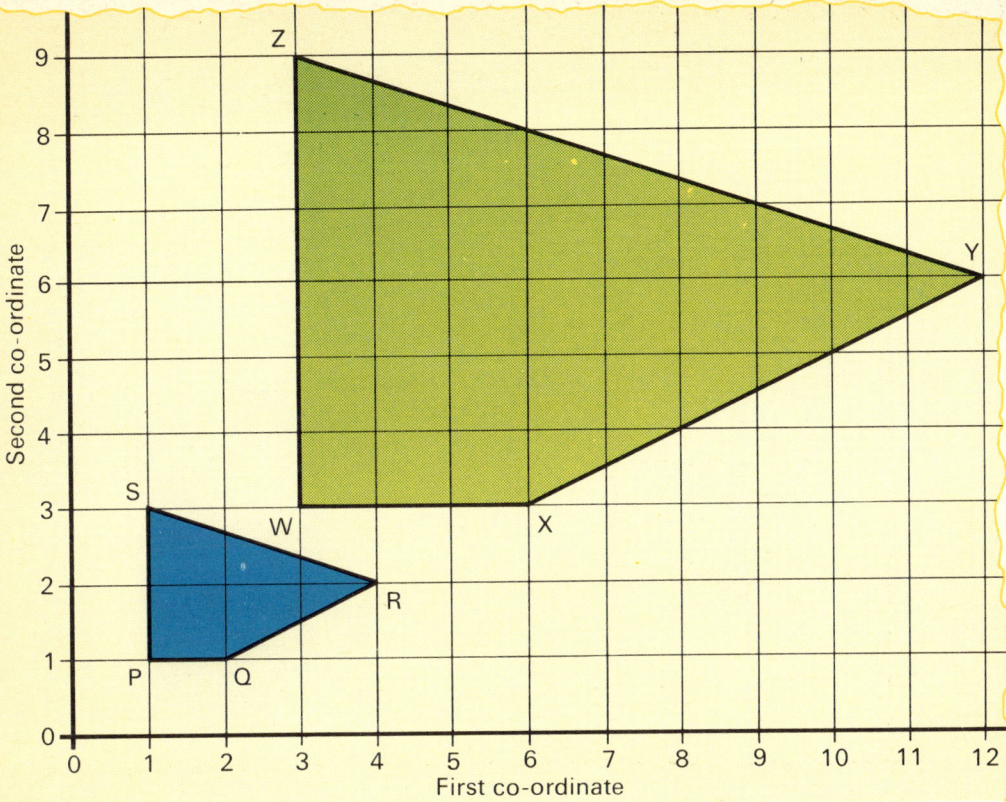

A   Look at the two shapes above. What can you say about them?
     What can you say about the lengths of: 1. PS and WZ?
     2. PQ and WX?

B   Write down the co-ordinates of:
     1. P and W   2. Q and X   3. R and Y   4. S and Z
     What do you notice?

C   Multiply each of the co-ordinates of P, Q, R, and S by two.
     Mark these new points on graph paper (W/s 47) and join them.
     Compare this image with the two shapes above. What do you notice?

D   A triangle has vertices whose co-ordinates are (1, 2), (3, 1), and (2, 3).
     Draw this triangle on graph paper (W/s 47).
     Find the image of each vertex, if its co-ordinates are multiplied by:
     1. four   2. two
     Mark the image of each vertex on the graph paper and join them.
     What can you say about the three triangles?

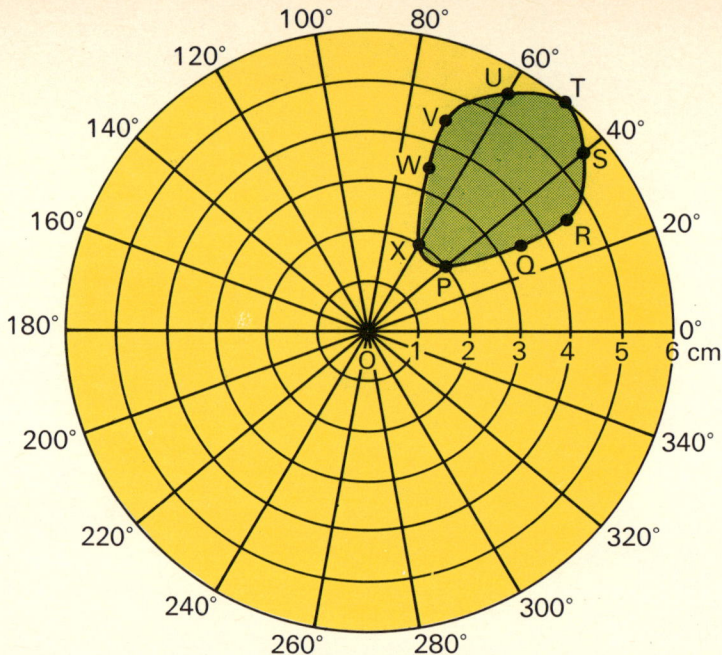

A   Write down the polar co-ordinates of the points P, Q, R, S, T, U, V, W, and X in the shape shown above.
     What are the co-ordinates of these points when the shape is rotated anticlockwise about O through: 1. 120°?   2. 240°?

B   Copy the diagram above (W/s 48) and draw the shape and its two images after the rotations in A.

C   Using O as the origin and the 0° line as shown on the right, find the polar co-ordinates of:
     1. R and X   2. S and Y
     3. T and Z
     What is the connection between each pair of co-ordinates?

D   What transformation could have been used to make XYZ the image of RST?

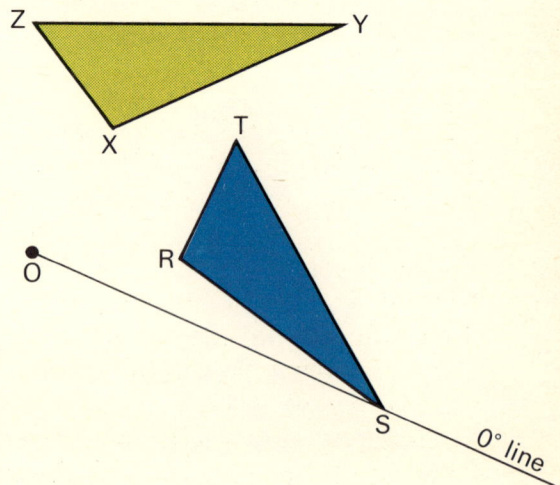

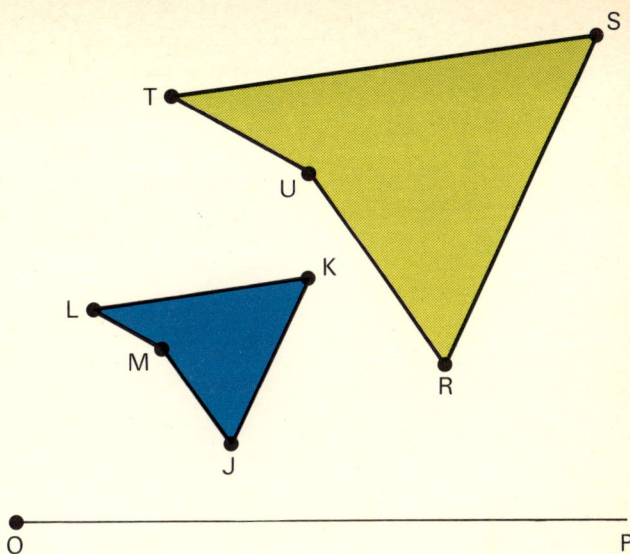

A   Look at the two shapes above.
What can you say about them?
What can you say about the lengths of:
1. OJ and OR?   2. OK and OS?   3. OL and OT?
4. OM and OU?

B   Using O as the origin and OP as the 0° line, find the polar
co-ordinates of:
1. J and R   2. K and S   3. L and T   4. M and U
What do you notice?

C   The polar co-ordinates of a
triangle XYZ are shown on
the right.
Using polar co-ordinates make
an accurate copy (W/s 49) of the
triangle. Measure its edges.

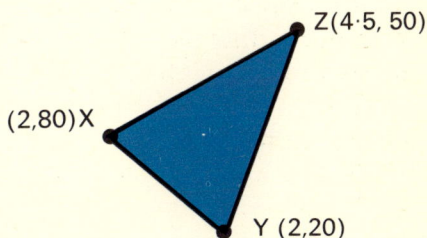

Z(4·5, 50)

(2,80)X

Y (2,20)

D   For each vertex of the
triangle drawn in C, treble
the distance from the origin.
Draw the new position of
each vertex (W/s 49).
Measure the edges of the new triangle.
Compare these with your results in C. What do you notice?

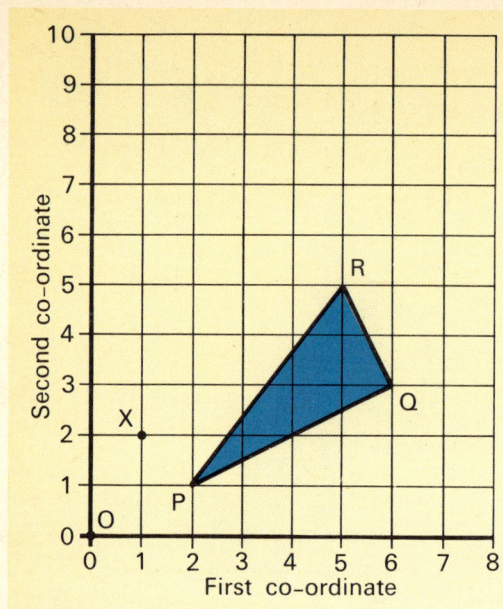

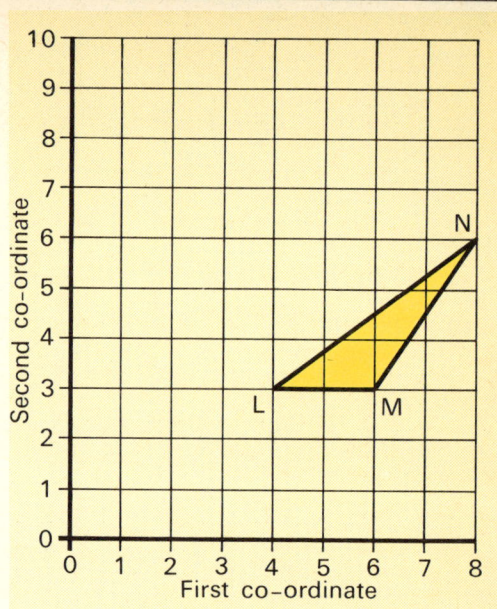

A | Look at the diagram on the left above.
The blue triangle is rotated in an anticlockwise direction about the point O. The image of the point P (2, 1) is the point X (1, 2). Find the co-ordinates of the images of the points Q and R.

B | Look at the diagram on the right above.
The yellow triangle is rotated 90° in an anticlockwise direction about the point L.
Find the co-ordinates of the images of the points L, M, and N.

C | Repeat B using the point M as the centre of rotation.

D | If the yellow triangle above is rotated 90° in a clockwise direction about the point N, find the co-ordinates of the images of the points L, M, and N.

E | Find the co-ordinates of the images of the points L, M, and N, if the yellow triangle above is rotated 180° about:
1. the point L   2. the point M

F | The yellow triangle above is rotated about the point (5, 5).
Find the co-ordinates of the images of the points L, M, and N for anticlockwise rotations of: 1. 90°   2. 180°   3. 270°   4. 360°.
What do you notice about your results?

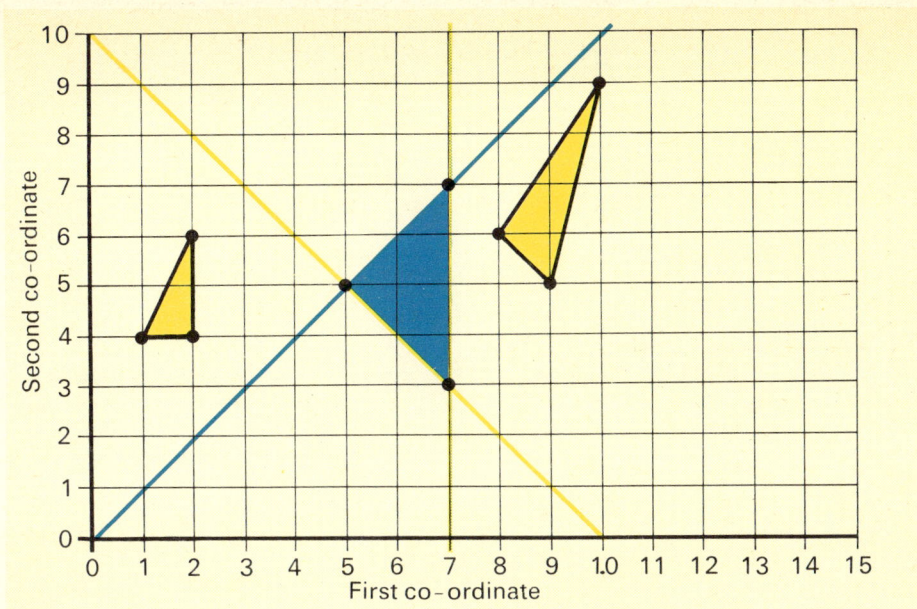

A Write down the co-ordinates of the vertices of the blue triangle.
Find the co-ordinates of the vertices of the blue triangle after it has
been flipped over:
1. the green line   2. the blue line   3. the yellow line

B Repeat A for: 1. the green triangle   2. the yellow triangle

C Use your results in A and B to make a table showing the co-ordinates
of the vertices of each triangle before and after it is flipped over the
blue line (e.g. the image of (1, 4) is (4, 1)). What do you notice about
your results? What would the image of (3, 10) be?

D Repeat C using   1. the green line   2. the yellow line

E Write down the co-ordinates of the image of the point (p, q)
after it has been flipped over
1. the green line   2. the blue line   3. the yellow line

F Draw the green triangle on squared paper (W/s 48).
Mark the blue line and the yellow line.
Flip the green triangle over the blue line and call its image I_B.
Flip the green triangle over the yellow line and call its image I_Y.
What transformation would make I_B the image of I_Y?

Review 2A

A

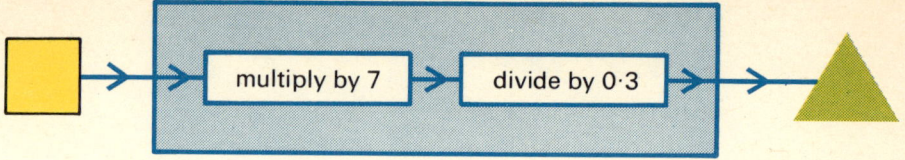

Find the number you will get in the green triangle if in the yellow box you put:

1. 6   2. 2·4   3. 7·2   4. 12   5. 0·9   6. 8   7. 4·1

B   Copy and complete the table about rectangles (W/s 49).

Length	5 cm	12 cm	7 cm		
Width	4 cm	1·4 cm		4 cm	0·9 cm
Area			42 cm²	21·6 cm²	4·23 cm²

C

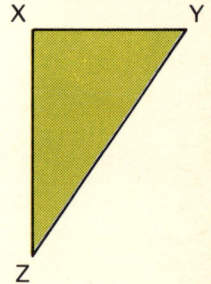

Explain how you could move
the triangle from its yellow position
to its green position.
Give as many different ways
as you can.

D   P = {3, 6, 9, 12, 15, 18, 21, 24}
Q = {4, 8, 12, 16, 20, 24, 28}
R = {5, 10, 15, 20, 25, 30}
Draw a Venn diagram to show P, Q, and R.

E   Using P, Q, and R from D write down:
1. $P \cap Q$ 2. $P \cap R$ 3. $Q \cap R$
What can you say about $P \cap Q \cap R$?

F   Which numbers are in P *but not* in: 1. Q?   2. R?

G On a sheet of graph paper (W/s 50) draw the quadrilateral whose vertices are the points (2, 2), (3, 2), (3, 4), and (1, 3).
Now multiply each of the co-ordinates of the four points by two. Mark the four points represented by these new co-ordinates.
Join the four points to form a quadrilateral. Write down all you can about the two quadrilaterals you have drawn.

H On squared paper (W/s 50) draw a graph to change marks out of 40 to percentages. A way in which the axes can be drawn and labelled is shown on the right.
Use your graph to find the percentage, for a mark out of 40, of:
1. 27   2. 33   3. 9   4. 21

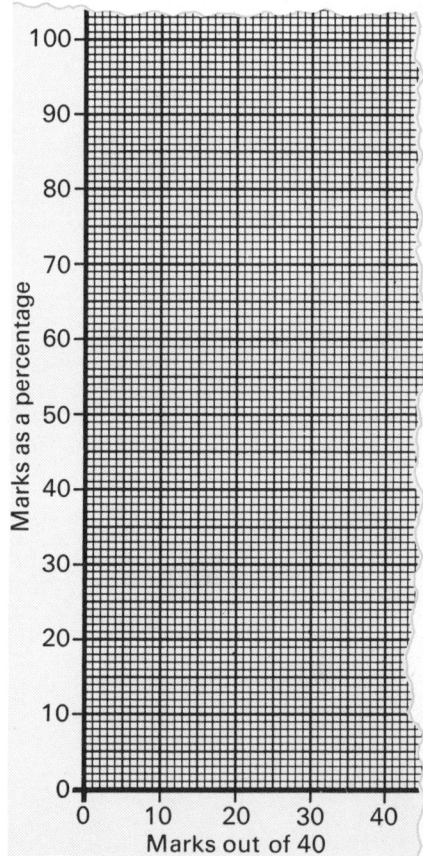

Marks as a percentage / Marks out of 40

I Write down as many pairs of whole numbers as you can which make a true statement for:
1. $\square + \triangle = 5$
2. $\square \times \triangle = 24$
3. $\square \div \triangle = 3$

J Write down the next number in the sequence:
1. 3, 6, 9, 12, 15, …
2. 20, 16, 12, 8, 4, …
3. 1, 4, 16, 64, …

K Work through the program. Say what calculation is done.

1. ①   3 → **L**
   ②   4 → **M**
   ③   5 → **N**
   ④ **L + M** → **W**
   ⑤ **W × N** → **R**
   ⑥   OUTPUT   **R**
   ⑦   STOP

2. ①   3 → **L**
   ②   4 → **M**
   ③   5 → **N**
   ④ **M × N** → **V**
   ⑤ **L + V** → **R**
   ⑥   OUTPUT   **R**
   ⑦   STOP

3. ①   3 → **L**
   ②   4 → **M**
   ③   5 → **N**
   ④ **L × N** → **W**
   ⑤ **M × N** → **V**
   ⑥ **W + V** → **R**
   ⑦   OUTPUT   **R**
   ⑧   STOP

Two of the programs give the same output. Can you say why?

Review 2B

A Find the height of the triangle:

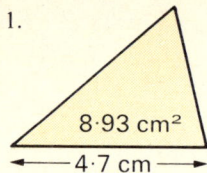

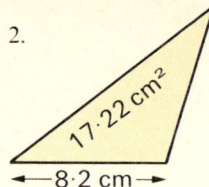

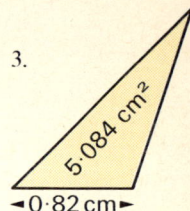

1.
8·93 cm²
←— 4·7 cm —→

2.
17·22 cm²
←— 8·2 cm —→

3.
5·084 cm²
◄0·82cm►

B On the left hand side of a sheet of
plain paper draw a regular hexagon (W/s 50).
Make a cardboard copy of your hexagon.
Show the position of the hexagon after:
a translation of 6 cm parallel to the top
edge of your paper; then a clockwise
half-turn rotation about P; then a clockwise
half-turn rotation about Q; and finally a
clockwise half-turn rotation about R.
Can you suggest other transformations which will move your
original hexagon to the same final position?

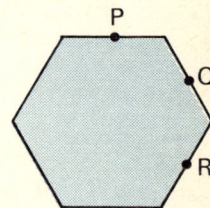

P
Q
R

C If P = {square numbers}, Q = {multiples of 4}, R = {multiples of 5}
write down the members of P, Q, and R for whole numbers
from 1 to 30. Use your results to find:
1. P ∩ Q   2. Q ∩ R   3. P ∩ R   4. P ∪ Q   5. P ∪ R   6. Q ∪ R
7. P ∩ Q ∩ R   8. P ∪ Q ∪ R   9. P \ Q   10. P \ R   11. Q \ R

D From {James, Brian, Peter, Rosalind, Anne} a boy and a girl
are chosen. Write down all the possible pairs (boy, girl).
What is the probability that Peter and Anne are chosen?
What is the probability that Peter and Anne are not chosen?

E Find the area of the yellow face of the shape.
Work through the program
shown on the right.
Can you explain why you
could use the program to
find the area of the face?
Find the volume of the
shape.

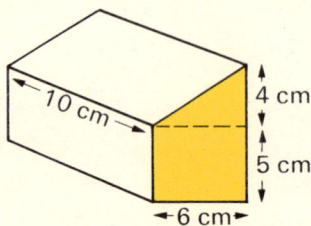

10 cm
4 cm
5 cm
◄6 cm►

①
②
③
④
⑤ B × L → R
⑥ B × H → W
⑦ W ÷ K → T
⑧ R + T → A
⑨ OUTPUT   A
⑩ STOP

2 → K
6 → B
5 → L
4 → H

Review 2C

A Find the area of the trapezium, shown on the right, when
$a = 10, b = 4, h = 3$.
Find the area in as many different ways as you can. (The dotted lines may help.)
If the area of the trapezium is A cm^2, write down an equation showing the relationship between A, a, b, and h.

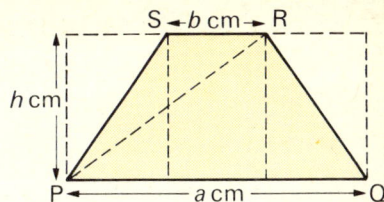

B

The relationship between x and y is shown by the arrow graph above. Write down an equation to show this relationship.
If x and y are members of {whole numbers from 1 to 12}, find the ordered pairs which belong to the solution set for $y = (x \times 4) - 3$.
Show this set of ordered pairs on a graph (W/s 51).
On the same graph show the set of ordered pairs for $y = 12 - x$.
What is the intersection of the two sets of ordered pairs?

C The co-ordinates of the three vertices of a triangle are (4, 2), (6, 1), and (7, 4). Find the co-ordinates of the vertices after the triangle has been rotated through 180° about the point (5, 3) (W/s 51).

D Eighteen children were asked to write down one of the numerals from 1 to 6.
The number of children writing down each of the numerals is shown by the pie chart on the right.
How many children wrote down each numeral?
If Peter is one of the children, what is the probability that he wrote down:
1. 4?   2. 1?   3. 5?
What is the probability that he wrote down an even number?

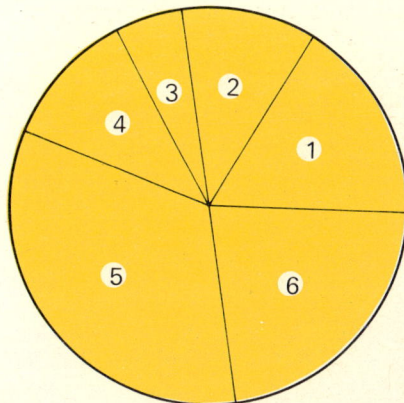

Circles and π

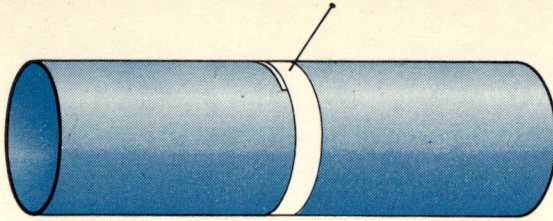

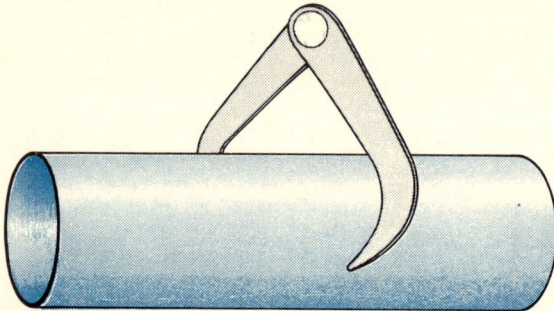

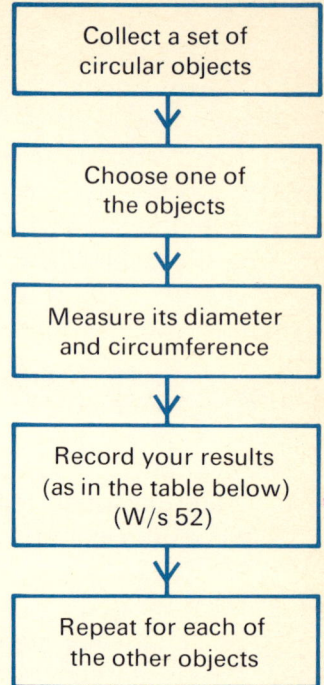

Collect a set of
circular objects

↓

Choose one of
the objects

↓

Measure its diameter
and circumference

↓

Record your results
(as in the table below)
(W/s 52)

↓

Repeat for each of
the other objects

Object	Circumference (C cm)	Diameter (D cm)	$C \div D$
tin	22·9 cm	7·3 cm	
plate	47·7 cm	15·2 cm	
clock			

A   Follow the instructions in the flow chart.
Do not complete the last column in the table.

B   What is the approximate value of $C \div D$ for:
1. the tin?   2. the plate?

C   Write down the approximate value of $C \div D$ for each of your set
of circular objects. What do you notice?

The circumference of a circle is just over three times the
diameter of the circle.

A   A bicycle wheel has a diameter
of 70 cm.
About how long is the
circumference of the wheel?

B   Each clock face of Big Ben has
a diameter of about 6·9 metres.
What is the approximate circumference
of each face?
About how far will the tip of the
minute hand move each minute?

C   A birthday cake has been cooked
in a tin with a diameter of 30 cm.
A paper frill has to be put round
the cake.
What length frill would you buy?

D   The diameter of a tin of baked beans
is 7·5 cm. Its height is 6 cm.
A label covers all the curved part
of the tin.
What shape is the label before it is
stuck on the tin?
What is the approximate area of
the label?

E   Choose some other circular objects. Measure the diameter of each
and then work out an approximate value for the circumference.
Check your approximate values by measuring.

F   Complete the fourth column of your table (on the opposite page)
(W/s 52). What do you notice about the values of $C \div D$?

G   Find the average of your set of values for $C \div D$.
Compare your average with those of your friends.
What do you notice?

More than 2000 years ago men had discovered a relationship
between the circumference and diameter of a circle. They used
it in designing their buildings.
At first they used the relationship:
        circumference = (diameter) × 3
They then used more exact relationships such as:
        circumference = (diameter) × $3\frac{10}{71}$

For centuries men tried to find the exact value of the 'just
over three' number—first as an ordinary fraction and then as
a decimal. But they did not succeed. They could not work out
the ordinary fraction and when they used decimals they found
that the decimal went on and on. It did not terminate.
For example, they found that:
        circumference = (diameter) × 3·141 592 653 589 793 238 46...
With a computer the number has now been found to hundreds and
even thousands of places of decimals.

To avoid using such complicated decimals, a symbol is used for
the number.
    The symbol is **π**. It is called **pi**.

Using π, the relationship is: circumference = (diameter) × π

Using C for circumference and D for diameter, the relationship
is: $C = D \times \pi$
This can be written as $C = \pi \times D$
To save time this is usually written as $\boldsymbol{C = \pi D}$.

For everyday calculations, 3·14 is often used for π ($\pi \simeq 3{\cdot}14$)
3·14 is approximately $3\frac{1}{7}$ ($\pi \simeq 3\frac{1}{7}$)

A Using 3·14 for π, calculate the circumference of a circle with a diameter of: 1. 12 cm   2. 20 cm   3. 5 m   4. 32 m   5. 9·4 cm   6. 13·6 cm   7. 23 mm   8. 67 m   9. 8·7 cm   10. 15·4 cm   11. 29·8 cm

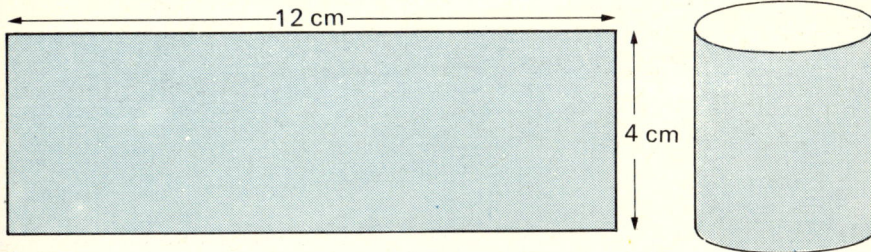

B Cut out a rectangular piece of paper, 12 cm by 4 cm.
Fold the paper to form an open cylinder.
Measure the diameter of the cylinder.
Calculate the diameter. Compare your two results.

The working of the calculation in B can be done in two ways.

(a) Using $\pi \simeq 3{\cdot}14$

Diameter (in cm) $= 12 \div 3{\cdot}14$

$= \dfrac{12}{3{\cdot}14}$

$= \dfrac{1200}{314}$

$= 3{\cdot}8$ (to one place of decimals)

$$
\begin{array}{r}
3{\cdot}82 \\
314)\overline{1200} \\
942 \\
\hline
2580 \\
2512 \\
\hline
680 \\
628 \\
\hline
\end{array}
$$

(b) Using $\pi \simeq 3\tfrac{1}{7}$

Diameter (in cm) $= 12 \div 3\tfrac{1}{7}$

$= 12 \div \dfrac{22}{7}$

$= \dfrac{12 \times 7}{22}$

$= \dfrac{84}{22}$

$= 3\tfrac{9}{11}$

$12 \div \tfrac{1}{7} = 12 \times 7$

$12 \div \dfrac{22}{7} = \dfrac{12 \times 7}{22}$

$$
\begin{array}{r}
3 \\
22)\overline{84} \\
66 \\
\hline
18 \\
\end{array}
$$

C Using (a) $\pi \simeq 3{\cdot}14$, (b) $\pi \simeq 3\tfrac{1}{7}$, find the diameter of a circle whose circumference is: 1. 88 cm   2. 50 cm   3. 14 m   4. 72 m

Indices

An electric light bulb can be either ON or OFF.

💡 shows that it is ON
💡 shows that it is OFF

One bulb is either 💡 or 💡

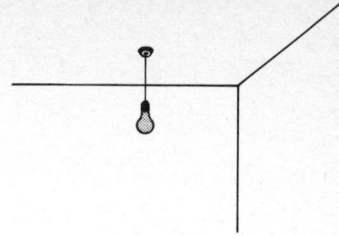

Two bulbs can be:

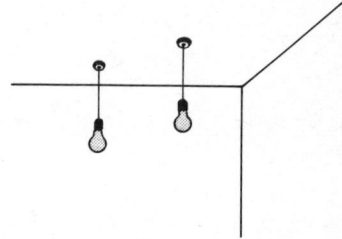

💡💡 💡💡 💡💡 💡💡

Three bulbs can be:

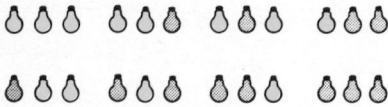

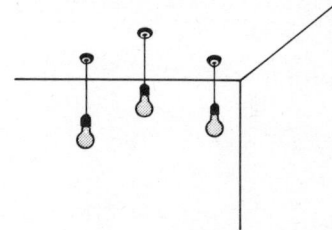

💡💡💡 💡💡💡 💡💡💡 💡💡💡

💡💡💡 💡💡💡 💡💡💡 💡💡💡

Four bulbs can be:

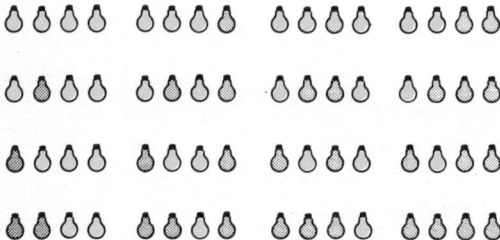

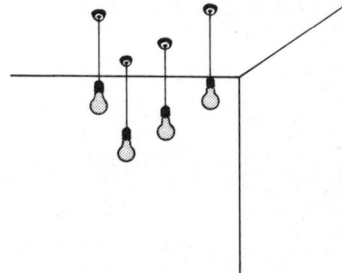

💡💡💡💡 💡💡💡💡 💡💡💡💡 💡💡💡💡

💡💡💡💡 💡💡💡💡 💡💡💡💡 💡💡💡💡

💡💡💡💡 💡💡💡💡 💡💡💡💡 💡💡💡💡

💡💡💡💡 💡💡💡💡 💡💡💡💡 💡💡💡💡

For a set of bulbs, the number of different ways in which the various bulbs can be ON or OFF is shown in the table below.

Number of bulbs in the set	Number of different ways	
1	2	$(2 \qquad\qquad)$
2	4	$(2 \times 2 \qquad)$
3	8	$(2 \times 2 \times 2 \quad)$
4	16	$(2 \times 2 \times 2 \times 2)$

A   Extend the above table as far as you can (W/s 53).

A   Look again at the set of bulbs on the opposite page.
     Explain in words (without working out the answer) how you
     would find the number of different ways the lights could be
     ON or OFF if there were:
     1. 10 bulbs   2. 12 bulbs   3. 15 bulbs   4. 20 bulbs

     For A1 you probably used the idea of getting the answer by the
     set of repeated multiplications:
$$2 \times 2 \times 2 \times 2 \times 2 \times 2 \times 2 \times 2 \times 2 \times 2$$
     For A2 you probably used:
$$2 \times 2 \times 2 \times 2 \times 2 \times 2 \times 2 \times 2 \times 2 \times 2 \times 2 \times 2$$

     There is a short way of writing repeated multiplications like these.

     $2 \times 2 \times 2 \times 2 \times 2 \times 2 \times 2 \times 2 \times 2 \times 2$ is written 2^{10}.

     $2 \times 2 \times 2 \times 2 \times 2 \times 2 \times 2 \times 2 \times 2 \times 2 \times 2 \times 2$ is written 2^{12}.

     In the same way the answers to A3 and A4 are written as 2^{15} and 2^{20}.

B   Write in the short way:
     1. $2 \times 2 \times 2 \times 2 \times 2 \times 2$ 2. $3 \times 3 \times 3 \times 3 \times 3$
     3. $5 \times 5 \times 5 \times 5$ 4. $9 \times 9 \times 9 \times 9 \times 9 \times 9 \times 9$
     5. 8×8 6. $7 \times 7 \times 7$

C   Write in full, as repeated multiplications:
     1. 2^7 2. 3^4 3. 5^3 4. 8^5 5. 2^9 6. 9^2 7. 10^2 8. 10^6

     In 2^7, the 7 is called the **index**.
     The index in 3^4 is the 4, whilst 5^3 has an index 3.
     Note: The plural of index is **indices**.

D   Find the value of:
     1. 2^4 2. 3^2 3. 3^4 4. 5^3 5. 10^2 6. 10^3 7. 10^4 8. 10^6

E   Find the index to make the statement true.
     1. $2^\square = 16$ 2. $3^\square = 27$ 3. $4^\square = 64$ 4. $2^\square = 32$ 5. $9^\square = 729$

F   Find the number to make the statement true.
     1. $\square^3 = 8$ 2. $\square^4 = 81$ 3. $\square^5 = 32$ 4. $\square^3 = 343$
     5. $\square^4 = 10\,000$

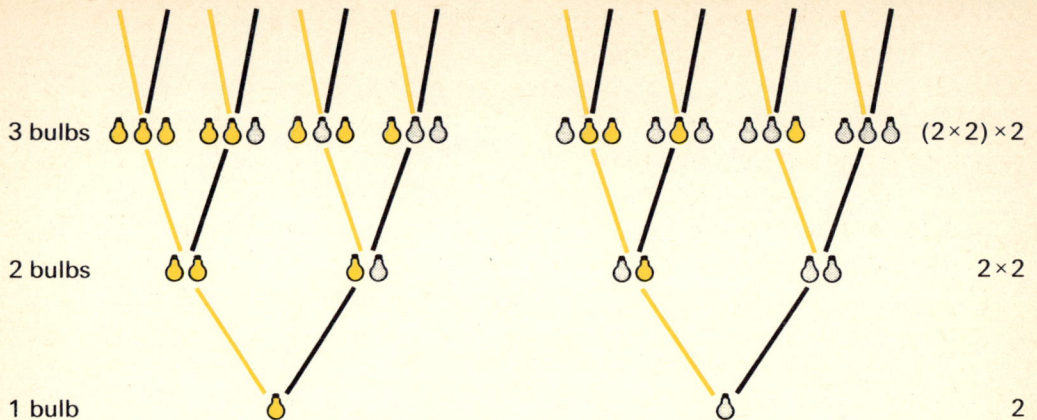

3 bulbs

2 bulbs

1 bulb

$(2 \times 2) \times 2$

2×2

2

The number of different ways in which a set of bulbs can be ON or OFF can be shown by using a tree, as started above.

A   On a large sheet of paper make a copy of the above tree and extend it to show the ways for 4 bulbs.

B   The result of a football match is a HOME win, an AWAY win, or a DRAW. Representing these as H, A, and D, a tree can be drawn to show the possible results of 1 match, 2 matches, 3 matches, etc. On a large sheet of paper (W/s 54) copy and extend the tree below.

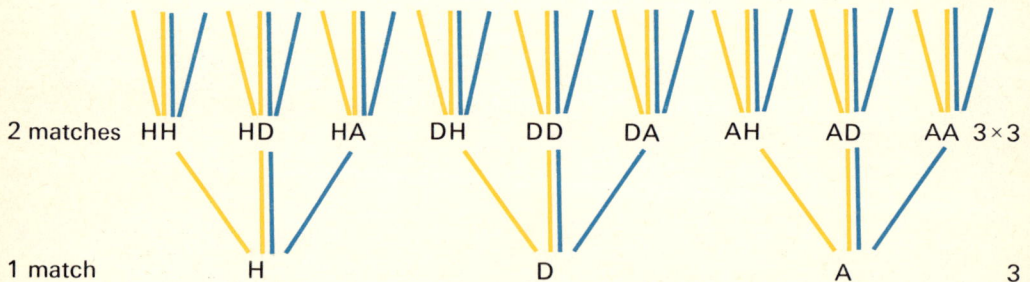

2 matches   HH   HD   HA   DH   DD   DA   AH   AD   AA 3×3

1 match   H   D   A   3

C   Using an index, write down the number of possible results of:
1. 3 matches   2. 4 matches   3. 8 matches   4. 12 matches

D   In a football competition the results of 6 matches have to be forecast. How many entries must you send in so as to be sure of correctly forecasting the six results?

2^4 is read as **two to the power four**.
10^6 is read as **ten to the power six**.

Look at the pattern formed
by the indices on the right.
In order to complete the
pattern, the number 2 is given
the index 1.

$$2 \times 2 \times 2 \times 2 \times 2 \times 2 = 2^6$$
$$2 \times 2 \times 2 \times 2 \times 2 = 2^5$$
$$2 \times 2 \times 2 \times 2 = 2^4$$
$$2 \times 2 \times 2 = 2^3$$
$$2 \times 2 = 2^2$$
$$2 = 2$$

$2 = 2^1$

A   Write in words: 1. 5^4 2. 7^6 3. 9^1 4. 20^5 5. 10^{10}

·B   Follow the flow chart shown
on the right.

C   Use the flow chart again but
replace 10 by:
1. 2   2. 3   3. 4   4. 5

Flow chart:

START

Write down 1

Multiply by 10.
Use an index to
show your answer

Have you reached 10^6? — No

Yes

Write down the
number which
10^6 represents

STOP

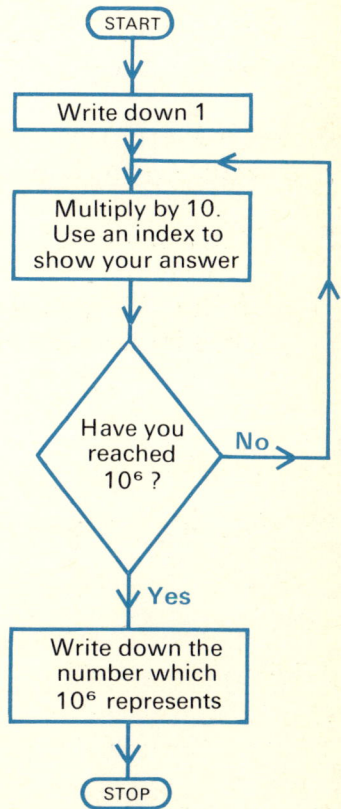

D   Work through the program below:

① $1 \rightarrow A$
② $1 \rightarrow P$
③ $0 \rightarrow C$
④ $10 \rightarrow N$
⑤ $P \times N \rightarrow P$
⑥ $C + A \rightarrow C$
⑦ IF $C < 6$ GO TO ⑤
⑧   OUTPUT P
⑨   STOP

E   Write, using an index:
1. $n \times n \times n \times n \times n \times n$ 2. $a \times a \times a \times a \times a$
3. $p \times p \times p \times p$ 4. $q \times q \times q \times q \times q \times q \times q$
5. $d \times d \times d \times d \times d \times d \times d \times d \times d \times d \times d \times d \times d$
6. t 7. $m \times m \times m \times m + m \times m \times m \times m$

A   Write, using an index, the
    area of a square whose edges are:
    1. 3 cm   2. 5 cm   3. 8 cm   4. 10 cm

B   Write, using an index, the
    volume of a cube whose edges are:
    1. 3 cm   2. 5 cm   3. 8 cm   4. 10 cm

The phrase **five squared** is often used for 5^2.

5×5 ➔ 5^2 ➔ five to the power two ➔ five squared.

The phrase **five cubed** is often used for 5^3.

$5 \times 5 \times 5$ ➔ 5^3 ➔ five to the power three ➔ five cubed.

C   Find the value of:
    1. seven squared   2. three cubed   3. eight squared
    4. ten cubed   5. six squared   6. a hundred cubed.

D   Write in words: 1. 8^2 2. 5^3 3. 7^3 4. 12^2 5. 24^3 6. 50^2

E   Find the value of:
    1. $1 \cdot 2^2$ 2. $2 \cdot 4^2$ 3. $4 \cdot 1^2$ 4. $1 \cdot 3^3$ 5. $2 \cdot 1^3$ 6. $8 \cdot 9^2$

F   Copy and complete:
    1. $1^2 =$ 2. $0 \cdot 1^2 =$ 3. $1 \cdot 1^2 =$
        $2^2 =$ $0 \cdot 2^2 =$ $1 \cdot 2^2 =$
        $3^2 =$ $0 \cdot 3^2 =$ $1 \cdot 3^2 =$
        $4^2 =$ $0 \cdot 4^2 =$ $1 \cdot 4^2 =$
        $5^2 =$ $0 \cdot 5^2 =$ $1 \cdot 5^2 =$
        $6^2 =$ $0 \cdot 6^2 =$ $1 \cdot 6^2 =$
        $7^2 =$ $0 \cdot 7^2 =$ $1 \cdot 7^2 =$
        $8^2 =$ $0 \cdot 8^2 =$ $1 \cdot 8^2 =$
        $9^2 =$ $0 \cdot 9^2 =$ $1 \cdot 9^2 =$
        $10^2 =$ $1 \cdot 0^2 =$ $2 \cdot 0^2 =$

A graph to show squares of numbers

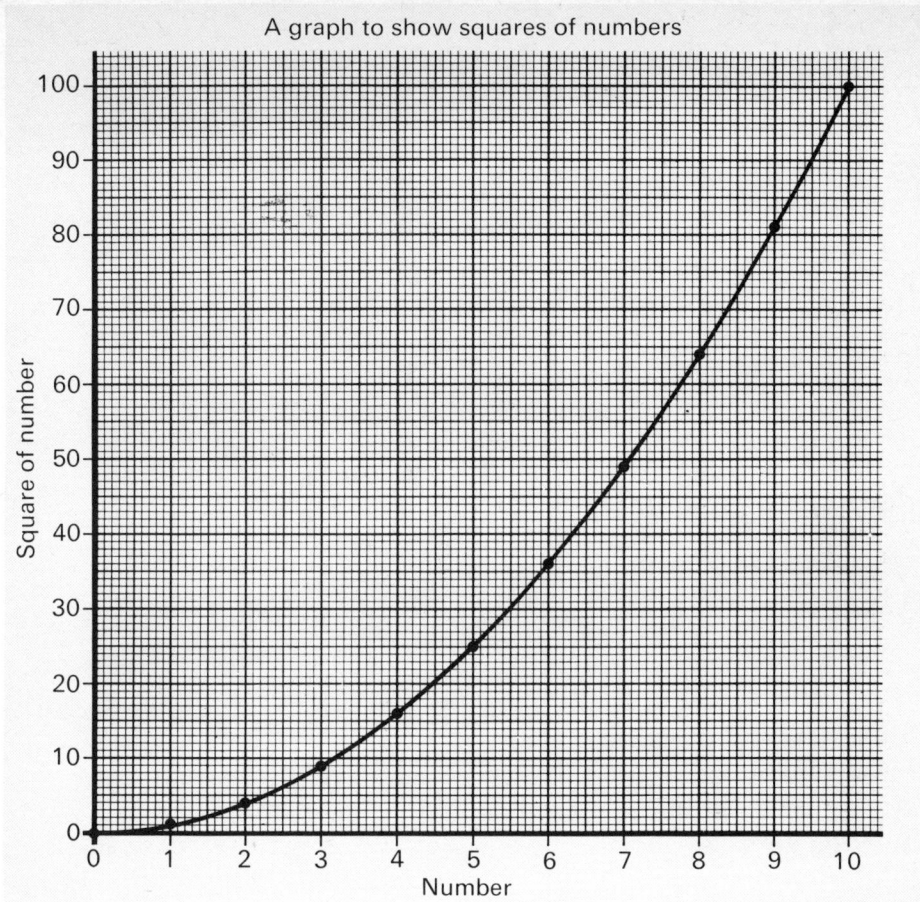

A   Use the above graph to find as accurately as possible:
1. 5^2 2. $1{\cdot}5^2$ 3. $2{\cdot}5^2$ 4. $4{\cdot}5^2$ 5. $8{\cdot}5^2$ 6. $7{\cdot}2^2$ 7. $8{\cdot}9^2$

B   Use the graph to find the number whose square is:
1. 36   2. 30   3. 55   4. 75   5. 19   6. 96   7. 84   8. 60

$7^2 = 49$ 49 is the **square** of 7.
              7 is the **square root** of 49. $7 = \sqrt{49}$

$1{\cdot}2^2 = 1{\cdot}44$ $1{\cdot}44$ is the square of $1{\cdot}2$.
            $1{\cdot}2$ is the square root of $1{\cdot}44$. $1{\cdot}2 = \sqrt{1{\cdot}44}$

C   Use the graph to find: 1. $\sqrt{81}$ 2. $\sqrt{69}$ 3. $\sqrt{54}$ 4. $\sqrt{27}$

Vectors

A   Use an ordered triple to record the number of stamps of each value in the 10p strip shown.

The contents of a 20p book includes the same stamps as the 10p strip, together with one more 4p stamp and two 3p stamps.

B   Use an ordered four to record the number of stamps of each value in a 20p book.

Recording the number of 4p, 3p, 2p, and 1p stamps in order, your answer to B would look like (2, 2, 2, 2).
The stamps in the 10p strip could also be recorded as an ordered four (1, 0, 2, 2). The 0 in this four means there are no 3p stamps in the strip.

Often, ordered pairs, triples and fours are written as **column vectors**. Three examples are shown on the right.

$$\begin{pmatrix} 1 \\ 2 \\ 2 \end{pmatrix} \quad \begin{pmatrix} 2 \\ 2 \\ 2 \\ 2 \end{pmatrix} \quad \begin{pmatrix} 1 \\ 0 \\ 2 \\ 2 \end{pmatrix}$$

C   Use an ordered four to record the number of stamps of each value in:
1. two 10p strips   2. two 20p books
3. one 10p strip and one 20p book

The stationery section of a Post Office sells special packets of parcel wrappings. The small packet has the contents shown below.

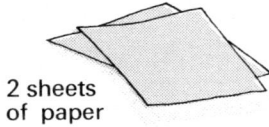

2 sheets of paper   2 pieces of string   5 sticky labels

A   Record the information above as a column vector.

The 'super' parcel wrappings packet has 5 sheets of brown paper, 6 pieces of string and 10 sticky labels.

B   Using the same order as you used in A, record the contents of the super packet as a column vector.

C   If you bought one packet of each kind, what single column vector would represent your purchase?

D   What single column vector would represent your purchase if you bought: 1. two super packets?   2. three small packets?
     3. two super packets and three small packets?

E   Using the column vector $\begin{pmatrix} sheets\ of\ paper \\ pieces\ of\ string \\ sticky\ labels \end{pmatrix}$ to represent a
     purchase of some packets, find how many small and super packets you need to buy to obtain:

1. $\begin{pmatrix} 4 \\ 4 \\ 10 \end{pmatrix}$ 2. $\begin{pmatrix} 9 \\ 10 \\ 20 \end{pmatrix}$ 3. $\begin{pmatrix} 10 \\ 12 \\ 20 \end{pmatrix}$ 4. $\begin{pmatrix} 11 \\ 12 \\ 25 \end{pmatrix}$ 5. $\begin{pmatrix} 8 \\ 8 \\ 20 \end{pmatrix}$

F   Using the same order as the column vector in E, record as a column vector, how much more a super packet has than a small packet.

G   What single column vector in E would represent your purchase if you bought: 1. p packets of each type?
     2. l super packets and m small packets?

The soldiers on the parade ground will move (*translate*) themselves two paces forward on the command given to them by the sergeant.

A   Copy the diagram on the right (W/s 55).
Draw the shape in its new position after a translation:
1. *move four squares right*
2. *move two squares upwards*
3. *move two squares right* followed by *one square upwards*.

The blue triangle is the image of the yellow triangle after a translation. This translation could be obtained by: *move to the right 2 cm,* followed by *move upwards 5 cm.*

A way of describing this translation is to use the single column vector:

$$\begin{pmatrix} 2 \\ 5 \end{pmatrix}$$

A column vector describing a translation is always used in the order:

$$\begin{pmatrix} distance\ moved\ to\ the\ right \\ distance\ move\ upwards \end{pmatrix}$$

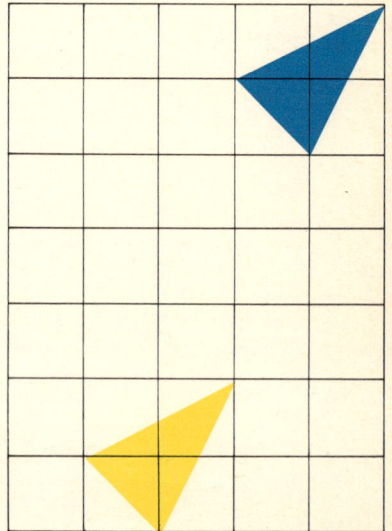

A   Measuring distance in centimetres, use a single column vector to
    describe the translation shown below which makes the blue shape
    the image of the yellow shape.

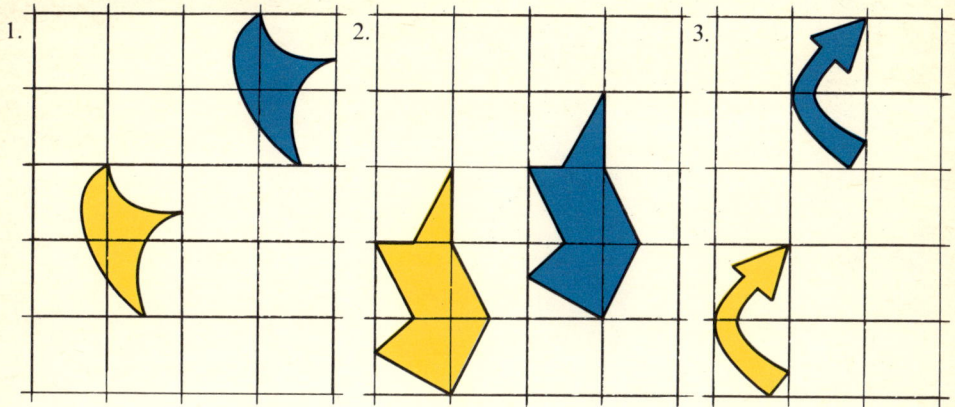

1.   2.   3.

B   Draw a simple shape in the bottom left hand corner of a sheet
    of squared paper (W/s 55).
    Draw the image of your shape after it has been moved by the
    translation described by the column vector:

    1. $\begin{pmatrix} 2 \\ 1 \end{pmatrix}$ 2. $\begin{pmatrix} 1 \\ 3 \end{pmatrix}$ 3. $\begin{pmatrix} 2 \\ 5 \end{pmatrix}$ 4. $\begin{pmatrix} 3 \\ 3 \end{pmatrix}$ 5. $\begin{pmatrix} 4 \\ 0 \end{pmatrix}$ 6. $\begin{pmatrix} 0 \\ 6 \end{pmatrix}$.

C   A knight in the game of chess
    can only move in certain ways.
    Some of these are shown on the
    right.
    What translation describes
    the move from:
    1. K to A? 2. A to B? 3. B to C?

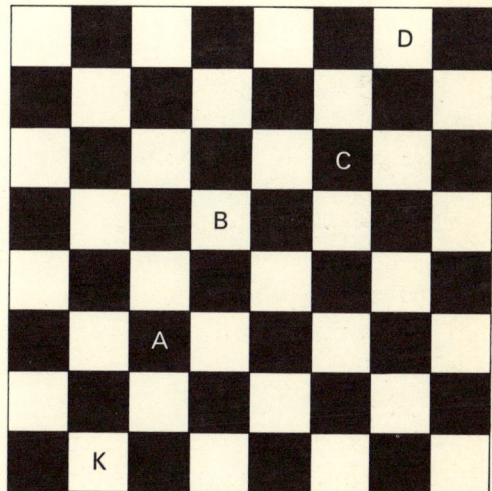

D   What single translation describes
    the knight's double move from
    K to B? What is the connection
    between this translation and the
    translations for the moves from
    K to A and from A to B?

E   Write down the translations for the moves from B to C and C to D.
    Can you use these two column vectors to write down the double
    move B to D?

A   Using the drawing on the right, find the column vector which describes the translation that makes:
1. P the image of the point (0, 0)
2. Q the image of P
3. Q the image of the point (0, 0)

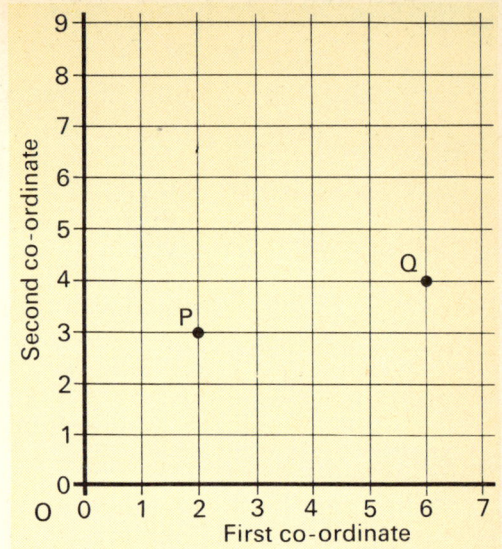

B   How could your answers to A1 and A2 be used to find the answer to A3?

C   The point (0, 0) is translated to a new position X, using the column vector $\begin{pmatrix} 3 \\ 2 \end{pmatrix}$.

What are the co-ordinates of X?

X is translated using $\begin{pmatrix} 4 \\ 6 \end{pmatrix}$ to the point Y.

What are the co-ordinates of Y?
Check your answer by drawing the points on graph paper (W/s 56).

D   Using the drawing on the right, describe the translation which makes
1. S the image of T
2. U the image of V
3. L the image of M
4. J the image of K
What do you notice?
What can you say about the lines TS, VU, ML, KJ?

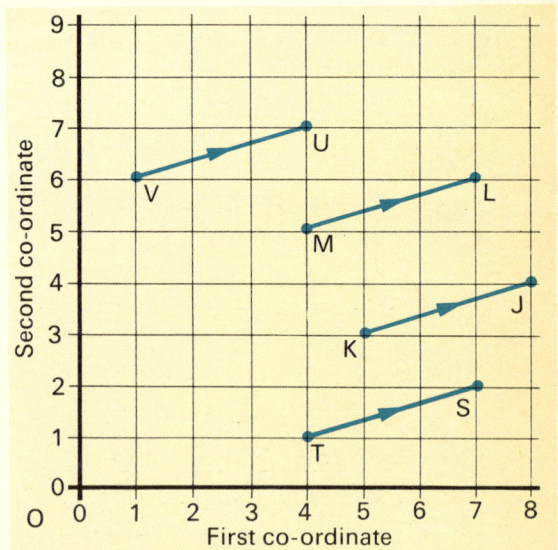

E   The point O(0, 0) is translated to A, B, C, and D by using the vectors
$\begin{pmatrix} 1 \\ 2 \end{pmatrix}$, $\begin{pmatrix} 2 \\ 4 \end{pmatrix}$, $\begin{pmatrix} 3 \\ 6 \end{pmatrix}$, and $\begin{pmatrix} 4 \\ 8 \end{pmatrix}$.
Show these points on a graph (W/s 56).
What do you notice about O, A, B, C, and D?
What can you say about OA, OB, OC, and OD?

A | R has co-ordinates (4, 1). It is translated using $\binom{2}{3}$ followed by $\binom{3}{1}$. What are the co-ordinates of the final image point? What single translation could replace the two translations?

B | A point S with co-ordinates (2, 3) is translated using $\binom{3}{2}$ followed by $\binom{4}{2}$ followed by $\binom{2}{5}$. What are the co-ordinates of the final image point? What single translation could replace the three translations?

C | Starting at the point (1, 3), how many translations using $\binom{2}{1}$, would you need to reach the point (7, 6)?

D | Look at the diagram on the right. What are the co-ordinates of the point (0, 0) after the translation:

1. $\binom{1}{2}$?   2. $\binom{3}{1}$?

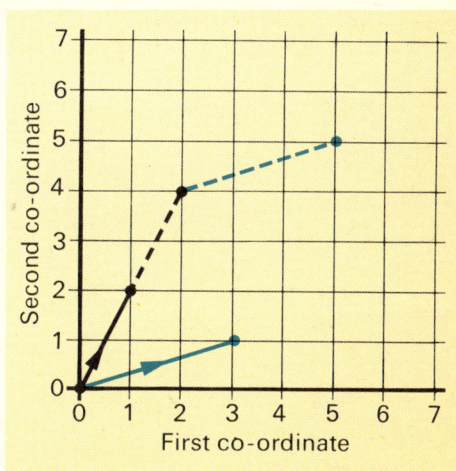

E | What are the co-ordinates of the point (0, 0) after it has been translated using $\binom{1}{2}$ followed by $\binom{1}{2}$ followed by $\binom{3}{1}$?

F | Using $\binom{1}{2}$ and $\binom{3}{1}$, find to what points on the diagram (0, 0) can be translated. In each case state the image point and how many times each translation is used.

G | Show ways of translating (2, 3) to (7, 8) using $\binom{1}{2}$ and $\binom{3}{1}$; (W/s 57).

H | Four types of sweet pack contain the numbers of different sweets shown in the table. Choose the number and type of each pack to give you exactly ten of each sweet.

	Type 1	Type 2	Type 3	Type 4
Chocolate	2	0	1	5
Crunchie	1	2	2	4
Polo	3	1	3	0
Milky Way	1	3	2	3
Fruit gums	0	4	3	3

Negative numbers

$3+4$ $5-2$ 2×6 $4\div8$ 0×7

$2\div7$

6×2 $8\div4$ $7+0$ $5+1$

$2-5$

$1+5$ $1\div5$ $4+3$ $1-5$

$5\div1$ $7-1$

$7\div2$ 1×5 7×0 5×1

$5-1$ $10-4$ $1-7$ $4-10$ $0+7$

A   To which of the examples shown above can you *not* give an answer?
Write them down. What do you notice?

B   Using whole numbers and fractions do you think it is possible
to give an answer to: 1. any addition?   2. any subtraction?
3. any multiplication?   4. any division?

C   Write down some subtractions to which you cannot give an
answer.

D   Here are some subtractions shown on a number line.
To which of them can you give the answer?

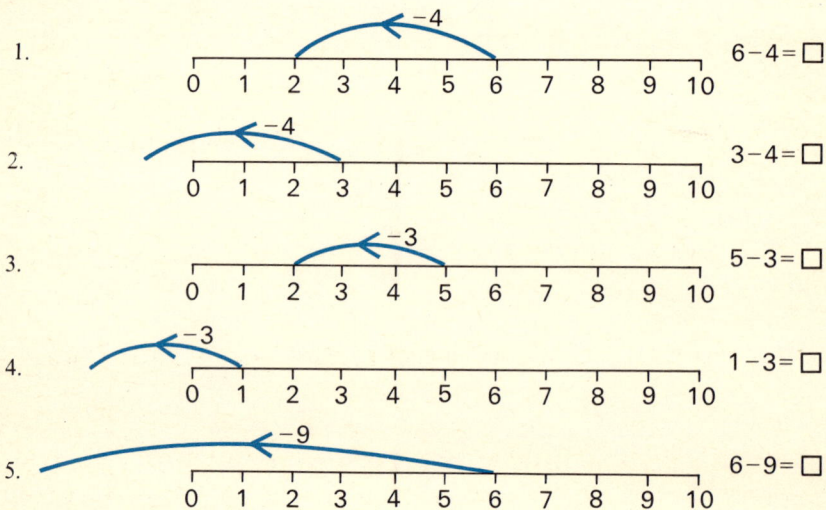

1. $6-4=\square$

2. $3-4=\square$

3. $5-3=\square$

4. $1-3=\square$

5. $6-9=\square$

A Look at the subtractions shown below. Each arrowed line goes to a point to the left of zero.
What can you say about the distance of each of these points from zero?

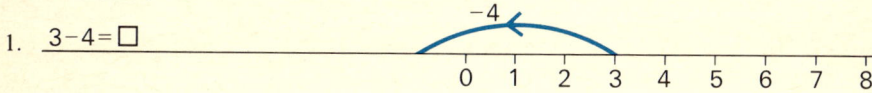

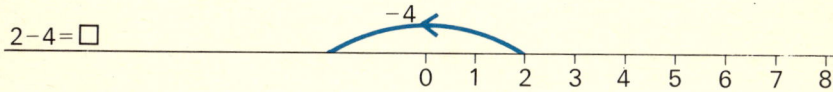

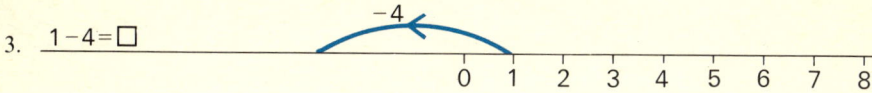

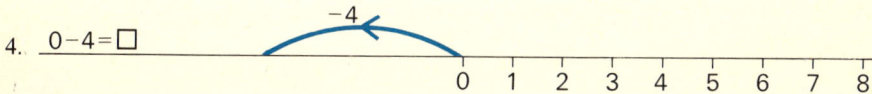

1. $3-4=\square$

2. $2-4=\square$

3. $1-4=\square$

4. $0-4=\square$

B If the points to the left of zero were numbered it would be possible to give an answer to each of the subtractions.
Could you number the points 1, 2, 3, 4, etc.? What answers would you then get to the subtractions? Would these be correct?
Suggest ways of numbering the points.

One way of numbering the points is shown below.

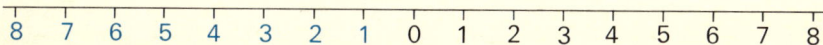

Using these blue numbers, the answers to the subtractions are:
$3 - 4 = 1$ $2 - 4 = 2$ $1 - 4 = 3$ $0 - 4 = 4$

Another way of numbering the points is to link the numbers with subtractions from zero. For example:

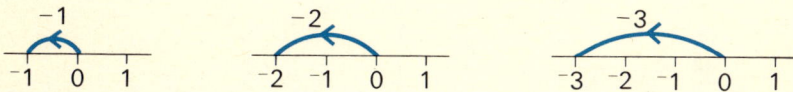

The numbers $^-1$, $^-2$, $^-3$, are read as **negative one, negative two,** and **negative three**.

Using negative numbers the answers to the subtractions above are:
$3 - 4 = {}^-1$ $2 - 4 = {}^-2$ $1 - 4 = {}^-3$ $0 - 4 = {}^-4$

Using negative numbers an extended number line is shown below.

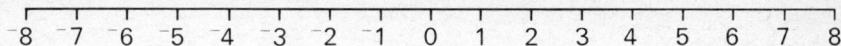

$$\begin{array}{cccccccccccccccccc} \text{-}8 & \text{-}7 & \text{-}6 & \text{-}5 & \text{-}4 & \text{-}3 & \text{-}2 & \text{-}1 & 0 & 1 & 2 & 3 & 4 & 5 & 6 & 7 & 8 \end{array}$$

A Use an arrow on a number line (W/s 58) to show the subtraction.
Write down the answer.
1. $4 - 5$ 2. $2 - 5$ 3. $5 - 7$ 4. $1 - 8$ 5. $6 - 6$ 6. $0 - 3$

B Write down the answer.
1. $6 - 8$ 2. $5 - 9$ 3. $0 - 6$ 4. $9 - 10$ 5. $7 - 12$ 6. $1 - 12$

{... ‾8, ‾7, ‾6, ‾5, ‾4, ‾3, ‾2, ‾1} is the set of negative whole numbers.

{1, 2, 3, 4, 5, 6, 7, 8, ...} is the set of positive whole numbers.

When the two sets are combined and zero is included, the set
of **integers** is formed.

{... ‾8, ‾7, ‾6, ‾5, ‾4, ‾3, ‾2, ‾1, 0, 1, 2, 3, 4, 5, 6, 7, 8, ...}

C Using integers, write down the co-ordinates of each marked point.

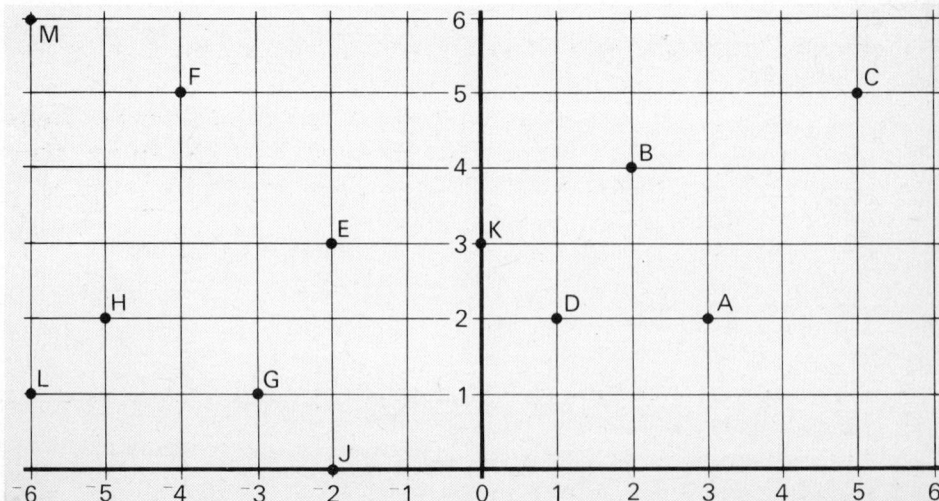

Before you go on to use the integers it may be helpful to write down the various properties of the positive integers which you now use, for example:

$$2 + 3 = 3 + 2 \qquad 4 \times 5 = 5 \times 4$$
$$7 + 0 = 7 \qquad 8 \times 1 = 8 \qquad 7 \times 0 = 0$$
$$0 + 7 = 7 \qquad 1 \times 8 = 8 \qquad 0 \times 7 = 0$$
$$(5 + 3) + 4 = 5 + (3 + 4) \qquad (2 \times 3) \times 4 = 2 \times (3 \times 4)$$

$$5\,(3 + 8) = (5 \times 3) + (5 \times 8)$$

It may also be helpful to look again at the number patterns for addition and multiplication.

+	0	1	2	3	4	5	6	7	8	9
0	0	1	2	3	4	5	6	7	8	9
1	1	2	3	4	5	6	7	8	9	10
2	2	3	4	5	6	7	8	9	10	11
3	3	4	5	6	7	8	9	10	11	12
4	4	5	6	7	8	9	10	11	12	13
5	5	6	7	8	9	10	11	12	13	14
6	6	7	8	9	10	11	12	13	14	15
7	7	8	9	10	11	12	13	14	15	16
8	8	9	10	11	12	13	14	15	16	17
9	9	10	11	12	13	14	15	16	17	18

×	0	1	2	3	4	5	6	7	8	9
0	0	0	0	0	0	0	0	0	0	0
1	0	1	2	3	4	5	6	7	8	9
2	0	2	4	6	8	10	12	14	16	18
3	0	3	6	9	12	15	18	21	24	27
4	0	4	8	12	16	20	24	28	32	36
5	0	5	10	15	20	25	30	35	40	45
6	0	6	12	18	24	30	36	42	48	54
7	0	7	14	21	28	35	42	49	56	63
8	0	8	16	24	32	40	48	56	64	72
9	0	9	18	27	36	45	54	63	72	81

These properties and patterns provide a starting point when you begin to use the integers in calculations. The new numbers— the negative integers—have no properties, except their positions on a number line. To make it easier to work with positive and negative integers at the same time, the whole set of integers is given the same properties as the positive integers.

For example: as $2 + 3 = 3 + 2$ so $^-4 + 2 = 2 + {}^-4$
as $7 + 0 = 7$ so $^-8 + 0 = {}^-8$
as $8 \times 1 = 8$ so $^-5 \times 1 = {}^-5$

A You know how to add two positive integers. For example, 5 and 9. Can you think of how you might add a positive integer and a negative integer, for example, 4 and $^-7$?
Can you think of how you might add two negative integers, for example, $^-3$ and $^-5$?

A   Write down the addition shown on the number line.

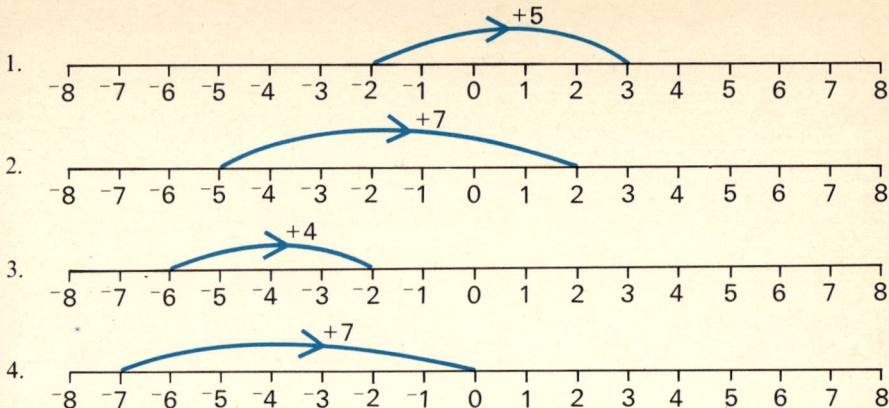

1.

$$-8 \quad -7 \quad -6 \quad -5 \quad -4 \quad -3 \quad -2 \quad -1 \quad 0 \quad 1 \quad 2 \quad 3 \quad 4 \quad 5 \quad 6 \quad 7 \quad 8$$

(+5)

2.

$$-8 \quad -7 \quad -6 \quad -5 \quad -4 \quad -3 \quad -2 \quad -1 \quad 0 \quad 1 \quad 2 \quad 3 \quad 4 \quad 5 \quad 6 \quad 7 \quad 8$$

(+7)

3.

$$-8 \quad -7 \quad -6 \quad -5 \quad -4 \quad -3 \quad -2 \quad -1 \quad 0 \quad 1 \quad 2 \quad 3 \quad 4 \quad 5 \quad 6 \quad 7 \quad 8$$

(+4)

4.

$$-8 \quad -7 \quad -6 \quad -5 \quad -4 \quad -3 \quad -2 \quad -1 \quad 0 \quad 1 \quad 2 \quad 3 \quad 4 \quad 5 \quad 6 \quad 7 \quad 8$$

(+7)

B   The additions in A are:
1. $^-2 + 5 = 3$ 2. $^-5 + 7 = 2$ 3. $^-6 + 4 = ^-2$ 4. $^-7 + 7 = 0.$
Using the property that $2 + 3 = 3 + 2$, $^-4 + 2 = 2 + {}^-4$, etc. and the results of A, write down the value of:
1. $5 + {}^-2$ 2. $7 + {}^-5$ 3. $4 + {}^-6$ 4. $7 + {}^-7$

C   Use a number line (W/s 59) to show the subtraction:
1. $5 - 2$ 2. $7 - 5$ 3. $4 - 6$ 4. $7 - 7$

D   Here are the results for A, B, and C:

$$^-2 + \ 5 = 3 \qquad ^-5 + \ 7 = 2 \qquad ^-6 + \ 4 = ^-2 \qquad ^-7 + \ 7 = 0$$
$$5 + {}^-2 = 3 \qquad 7 + {}^-5 = 2 \qquad 4 + {}^-6 = ^-2 \qquad 7 + {}^-7 = 0$$
$$5 - \ 2 = 3 \qquad 7 - \ 5 = 2 \qquad 4 - \ 6 = ^-2 \qquad 7 - \ 7 = 0$$

What do you notice?
Can you see a quick way of finding the answer when you add a negative integer and a positive integer?

E   Copy and complete the addition:
1. $6 + {}^-2$ 2. $8 + {}^-5$ 3. $9 + {}^-1$ 4. $10 + {}^-6$ 5. $8 + {}^-8$
6. $^-4 + \ 7$ 7. $^-2 + \ 9$ 8. $^-6 + \ 4$ 9. $^-3 + \ 1$ 10. $^-7 + \ 7$
11. $20 + {}^-2$ 12. $7 + {}^-18$ 13. $40 + {}^-50$ 14. $^-30 + \ 8$ 15. $^-45 + 29$

F   To which of these additions can you not find the answer?

$6 + {}^-1$ $^-2 + {}^-3$ $^-3 + 10$ $^-5 + {}^-5$ $8 + {}^-9$

$^-8 + {}^-8$ $^-7 + 0$ $1 + {}^-9$ $^-1 + {}^-10$ $^-8 + 8$

One way of finding the answer to the addition of two negative integers is to make use of the pattern in an addition matrix.

Second number

+	5	4	3	2	1	0	⁻1	⁻2	⁻3	⁻4	⁻5
5	10	9	8	7	6	5	4	3	2	1	0
4	9	8	7	6	5	4	3	2	1	0	⁻1
3	8	7	6	5	4	3	2	1	0	⁻1	⁻2
2	7	6	5	4	3	2	1	0	⁻1	⁻2	⁻3
1	6	5	4	3	2	1	0	⁻1	⁻2	⁻3	⁻4
0	5	4	3	2	1	0	⁻1	⁻2	⁻3	⁻4	⁻5
⁻1	4	3	2	1	0	⁻1					
⁻2	3	2	1	0	⁻1	⁻2					
⁻3	2	1	0	⁻1	⁻2	⁻3					
⁻4	1	0	⁻1	⁻2	⁻3	⁻4					
⁻5	0	⁻1	⁻2	⁻3	⁻4	⁻5					

(First number — left-hand column)

A   The matrix above, for the integers 5 to ⁻5, shows the additions which you can already do. Check that these are correct.

B   Look at the number patterns in the matrix.
Complete the matrix (W/s 60) by using the number patterns.

C   Use your completed matrix to find the answer to:
1. ⁻1 + ⁻2   2. ⁻3 + ⁻4   3. ⁻1 + ⁻5   4. ⁻4 + ⁻4   5. ⁻2 + ⁻5
What do you notice? Can you suggest a quick way of adding two negative integers?

D   Write down the answer:
1. ⁻6 + ⁻5   2. ⁻4 + ⁻8   3. ⁻7 + ⁻6   4. ⁻10 + ⁻2
5. ⁻9 + ⁻8   6. ⁻15 + ⁻20   7. ⁻19 + ⁻19   8. ⁻40 + ⁻60

Second number

$-$	5	4	3	2	1	0	$^-1$	$^-2$	$^-3$	$^-4$	$^-5$
5		1	2								
4											
3			0								
$^-2$											
1											
0				$^-2$							
$^-1$		$^-5$									
$^-2$											
$^-3$											
$^-4$				$^-6$							
$^-5$											

First number

A   A subtraction matrix has been started above.
Copy and complete as much of it as you can (W/s 61).

B   Did you manage to complete the whole of the matrix in A?
If not, did you think of extending the patterns formed by the
numbers in the rows and columns (as you did for addition)?
Try doing this.
Can you now complete the whole matrix?

C   Look at your completed matrix.
Write down any interesting results which you notice.

Second number

−	5	4	3	2	1	0	⁻1	⁻2	⁻3	⁻4	⁻5
5	0	1	2	3	4	5	6	7	8	9	10
4	⁻1	0	1	2	3	4	5	6	7	8	9
3	⁻2	⁻1	0	1	2	3	4	5	6	7	8
2	⁻3	⁻2	⁻1	0	1	2	3	4	5	6	7
1	⁻4	⁻3	⁻2	⁻1	0	1	2	3	4	5	6
0	⁻5	⁻4	⁻3	⁻2	⁻1	0	1	2	3	4	5
⁻1	⁻6	⁻5	⁻4	⁻3	⁻2	⁻1	0	1	2	3	4
⁻2	⁻7	⁻6	⁻5	⁻4	⁻3	⁻2	⁻1	0	1	2	3
⁻3	⁻8	⁻7	⁻6	⁻5	⁻4	⁻3	⁻2	⁻1	0	1	2
⁻4	⁻9	⁻8	⁻7	⁻6	⁻5	⁻4	⁻3	⁻2	⁻1	0	1
⁻5	⁻10	⁻9	⁻8	⁻7	⁻6	⁻5	⁻4	⁻3	⁻2	⁻1	0

(First number down the left; Second number across the top.)

A subtraction matrix for the integers ⁻5 to 5 is shown above.
The part to the left of the dotted line was completed by using a
number line. The part to the right of the line was completed by
extending the number patterns of the left hand part.

A Use the matrix to find the answer:
1. $4 - {}^-2$ 2. $5 - {}^-4$ 3. $1 - {}^-1$ 4. $3 - {}^-5$ 5. $0 - {}^-3$ 6. $0 - {}^-2$
What do you notice?

B Use the matrix to find the answer:
1. ${}^-2 - {}^-5$ 2. ${}^-1 - {}^-4$ 3. ${}^-3 - {}^-5$ 4. ${}^-4 - {}^-4$ 5. ${}^-1 - {}^-1$ 6. $0 - {}^-5$
7. ${}^-4 - {}^-2$ 8. ${}^-5 - {}^-3$ 9. ${}^-2 - {}^-1$ 10. ${}^-3 - {}^-3$ 11. ${}^-4 - {}^-3$ 12. $0 - {}^-1$

C Write down the answers to the pair of examples:
1. $4 - {}^-2$ 2. $5 - {}^-1$ 3. ${}^-3 - {}^-4$ 4. ${}^-4 - {}^-3$ 5. ${}^-2 - {}^-5$ 6. $0 - {}^-4$
   $4 + 2$ $5 + 1$ ${}^-3 + 4$ ${}^-4 + 3$ ${}^-2 + 5$ $0 + 4$

D From A, B, and C do you agree that the subtraction of a negative
   integer gives the same result as the addition of the corresponding
   positive integer?

E 1. $7 - {}^-4$ 2. $12 - {}^-3$ 3. ${}^-2 - {}^-10$ 4. ${}^-5 - {}^-15$ 5. ${}^-8 - {}^-3$ 6. ${}^-20 - {}^-4$

Solution sets

A The number line above shows all the integers from 0 to 16.
Describe in your own words the subset of numbers coloured:
1. yellow (*y*) 2. blue (*b*) 3. green (*g*)

B Say which of the subsets in A is the solution set for:
1. $\square = 8$ 2. $\square > 10$ 3. $\square < 6$

C Show on a number line for integers from 0 to 16 (W/s 62),
the solution set for: 1. $\square = 9$ 2. $\square < 4$ 3. $\square > 12$

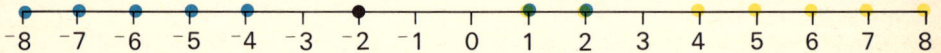

D The number line above shows all the integers from $^-8$ to 8.
Describe in your own words the subset of numbers coloured:
1. yellow 2. blue 3. black 4. green

E Say which of the subsets in D is the solution set for:
1. $\square > 3$ 2. $\square < ^-3$ 3. $\square + 2 = 0$

F Show on a number line for integers from $^-8$ to 8 (W/s 63), the
solution set for: 1. $\square - 6 = 0$ 2. $4 - \square = 0$ 3. $\square + 5 = 0$

G Show on a number line, for integers from $^-8$ to 8 (W/s 63), the
solution set for: 1. $\square < 3$ 2. $\square < 0$ 3. $\square < ^-2$

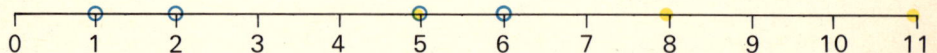

H The number line above shows all the integers from 0 to 11.
The solution set for $\square > 4$ is to be shown with yellow dots.
The solution set for $\square < 7$ is to be shown with blue circles.
Part of the diagram has been filled in. Copy and complete it (W/s 63).

I Use H to find the intersection of the solution sets for $\square > 4$ and
$\square < 7$.

0	1	2	3	4	5	6	7	8	9	10

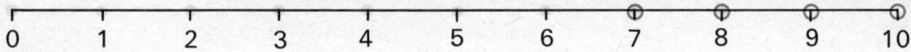

A A number line showing integers from 0 to 10 is shown above.
Describe in your own words the subset of numbers marked with:
1. yellow dots   2. blue circles   3. *both* dots *and* circles

The numbers on the number line marked in yellow are the subset of
numbers which make the statement ☐ < 9 true.
Those numbers marked with blue circles make the statement
☐ > 6 true.
The numbers which have *both* a yellow dot *and* a blue circle are
those numbers which make the statements ☐ < 9 *and* ☐ > 6 true.
{7, 8} is the intersection of the two solution sets.

B Find the numbers which belong to the solution set for ☐ < 9   *but*
not to the solution set for ☐ > 6.
How are these numbers marked on the number line above?

C Find the numbers which belong to the solution set for ☐ > 6 *but not*
to the solution set for ☐ < 9.
How are these numbers marked on the number line above?

D Using a number line, showing integers from 0 to 16 (W/s 62)
find the subset of numbers which make the statement true:
1. ☐ > 10 *and* ☐ < 13   2. ☐ < 12 *and* ☐ > 5   3. ☐ < 1
4. ☐ > 10 *but not* ☐ < 13   5. ☐ < 12 *but not* ☐ > 5

E Four brothers want to enter a competition.
To enter you must be:
I. a teenager but not old enough to drive a car
II. over 145 cm but under 170 cm in height.

	Age	Height
David	12	143 cm
Ian	13	147 cm
Andrew	16	172 cm
Charles	19	165 cm

Which of the brothers satisfy: 1. condition I?   2. condition II?
3. condition I *but not* condition II?   4. condition I *and* condition II?

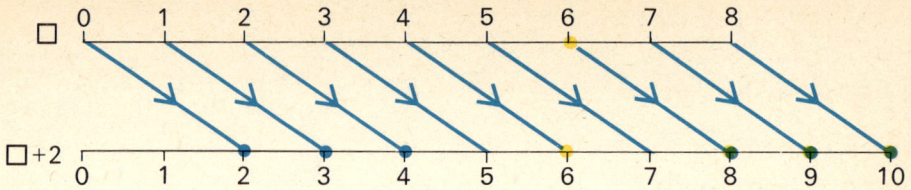

A   The arrow graph above shows □ → □ + 2 for integers
     (in the □) from 0 to 8.
     Write down the image of: 1. 3   2. 6   3. 8

B   Write down the number which has as its image: 1. 4   2. 7   3. 9

C   What number can be put in the □ to make the statement true:
     1. □ + 2 = 4?   2. □ + 2 = 7?   3. □ + 2 = 9?

D   Compare your answers to B and C. What do you notice?
     Use the arrow graph to solve □ + 2 = 8.

E   Use the arrow graph to help you to find the solution set for:
     1. □ + 2 < 5   2. □ + 2 > 7   3. □ + 2 = 6

F   Draw an arrow graph, like the one above, showing □ → □ + 5
     for the set of integers from 0 to 8 (W/s 64).
     Use your arrow graph to find the solution set for:
     1. □ + 5 = 8   2. □ + 5 = 11   3. □ + 5 < 7   4. □ + 5 > 9

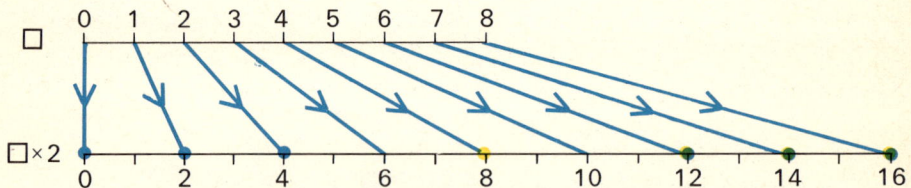

G   The arrow graph above shows □ → □ × 2 for integers 0 to 8.
     Use this arrow graph to solve: 1. □ × 2 = 6   2. □ × 2 = 14

H   Use the arrow graph, in G, to help you to find the solution set for:
     1. □ × 2 < 6   2. □ × 2 > 10   3. □ × 2 = 8

I   Draw an arrow graph, like the one in G, showing □ → □ × 3
     for the set of integers from 0 to 8 (W/s 64).
     Use this arrow graph to find the solution set for:
     1. □ × 3 = 15   2. □ × 3 = 21   3. □ × 3 < 12   4. □ × 3 > 9

A Use the set of integers from 0 to 20.
Find the subset of these which is the solution set for:
1. $(\square \times 3) + 2 > 40$ 2. $(\square \times 3) + 2 < 24$ 3. $(\square \times 3) + 2 = 35$

B Use the set of integers from 0 to 16.
Find the subset of these for which the statement is true:
1. $\square \times 5 < 50$ *and* $\square + 3 > 9$ 2. $\square \times 4 > 12$ *and* $\square - 3 < 4$

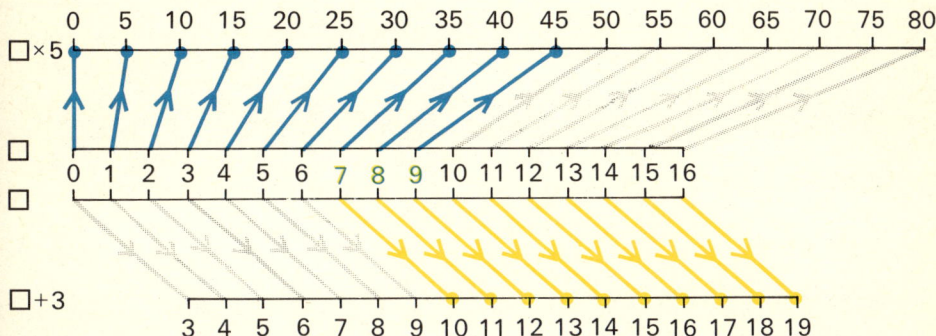

In the diagram above the blue dots on the top number line show where $(\square \times 5) < 50$. The yellow dots on the bottom number line show where $(\square + 3) > 9$. The only numbers that can be put in the $\square$ to make both statements true are 7, 8, and 9.

C Use the diagram above to find the solution set for:
1. $\square \times 5 < 50$ *but not* $\square + 3 > 9$ 2. $\square + 3 > 9$ *but not* $\square \times 5 < 50$

To save time you could use X to stand for the solution set for $\square \times 5 < 50$. In the same way you could use Y to stand for the solution set for $\square + 3 > 9$. C1 would be written as X *but not* Y.

D Use the set of integers from 0 to 20.
1. X stands for the solution set for $\square \times 4 > 50$.
   If x is a member of X, is it true to say $x \times 4 > 50$?
   Find the different numbers that x can be.

2. Y stands for the solution set for $\square + 10 < 25$.
   If y is a member of Y, is it true to say $y + 10 < 25$?
   Find the different numbers that y can be.

3. Find: (a) X *but not* Y   (b) Y *but not* X   (c) X ∩ Y

E Find x: 1. $x + 4 = 9$ 2. $7 - x = 2$ 3. $x - 8 = 13$
4. $(x \times 3) + 5 = 32$ 5. $(x \times 2) + 7 = 25$ 6. $(x \times 4) - 5 = 39$

A Look at the two sets of ordered pairs shown on the right. For each write down a □, △ statement.

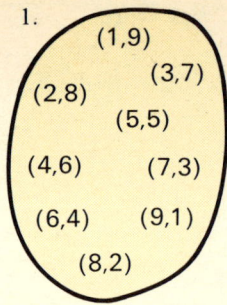

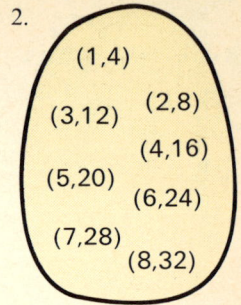

B Which ordered pair makes both statements in A true at the same time?

1.
(1,9)
(3,7)
(2,8)
(5,5)
(4,6) (7,3)
(6,4) (9,1)
(8,2)

2.
(1,4)
(3,12) (2,8)
(4,16)
(5,20)
(6,24)
(7,28)
(8,32)

C Look at the set of points shown in the graph on the right. Write down for this set a □, △ statement.

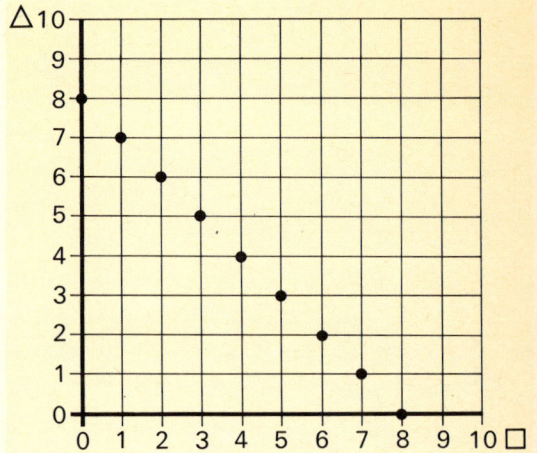

D Copy the graph on the right (W/s 65). On it mark the points for $\triangle = \square + 2$.

E Which ordered pair makes $\triangle = \square + 2$ and the statement in C true at the same time?

F Use a grid like the one in D (W/s 65). On it mark the set of ordered pairs for: 1. $\square + \triangle = 10$ 2. $\triangle = \square - 4$.

G Use your graph in F to help you to find the solution set for:
1. $\square + \triangle = 10$ *and* $\triangle = \square - 4$
2. $\square + \triangle = 10$ *but not* $\triangle = \square - 4$

H Look at the graph on the right. The black dots show $\square + \triangle = 6$. The blue dots show $\triangle = \square - 2$.
Copy the graph (W/s 66).
Mark in the points where:
1. $\square + \triangle < 6$ 2. $\triangle < \square - 2$
What can you say about these sets of points?

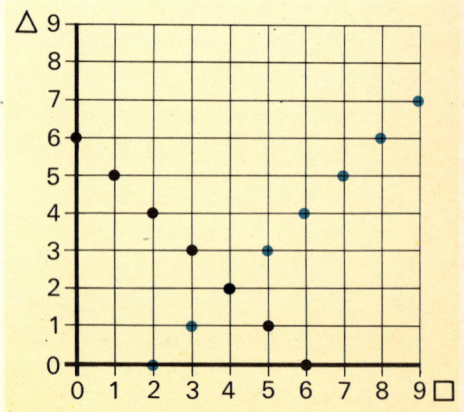

I Find the solution set for:
$\square + \triangle < 6$ *and* $\triangle < \square - 2$.

A seaside resort owns ten old buses (B) and eight old trams (T). These are used to ferry the holiday-makers around the town. Unfortunately, there are only twelve drivers for these vehicles.

A   Would it be reasonable to make the statement that, at any time, the number of buses and trams on the road $\leqslant 12$?

B   Write down the set of ordered pairs that make the statement $B + T \leqslant 12$ true. (4, 3) is one member of this set.

C   Copy and complete the graph shown on the right, marking in those points for which $B + T \leqslant 12$ (W/s 66).

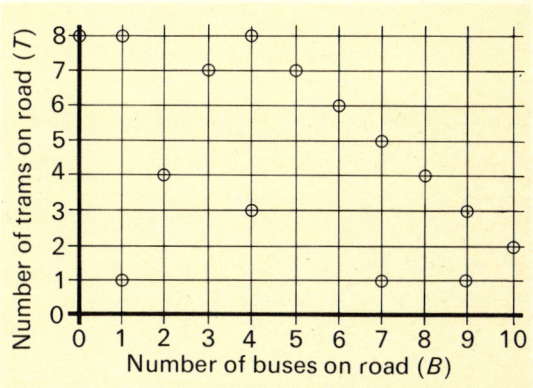

D   For which of the marked points on your graph in C is it true that:
1. $B > 6$ and $T > 2$?
2. $B > 3$ and $T > 5$?
3. $B > 7$ and $T > 3$?

E   If x and y are both members of {whole numbers from 1 to 12}, show on a graph (W/s 67) the ordered pairs which make the statement true:
1. $x + y = 9$ 2. $y = x + 5$ 3. $x \times 2 = y$

F   Use your graph in E to find the solution set for:
1. $x + y = 9$ and $y = x + 5$ 2. $x + y = 9$ and $x \times 2 = y$
3. $y = x + 5$ and $x \times 2 = y$

Using scales

A   Look at the drawings above.
Can you suggest ways of finding the height of:
1. a tree?   2. a block of flats?   3. a lamp post?
4. a tree on the far side of a river?

B   Pick out some tall objects which you can see from your classroom
or playing field.
Estimate the height of each of them.
Record your estimates. Compare them with your friends.
Now try finding the heights, using the methods you suggested
for A. Again, compare your results with those of your friends.

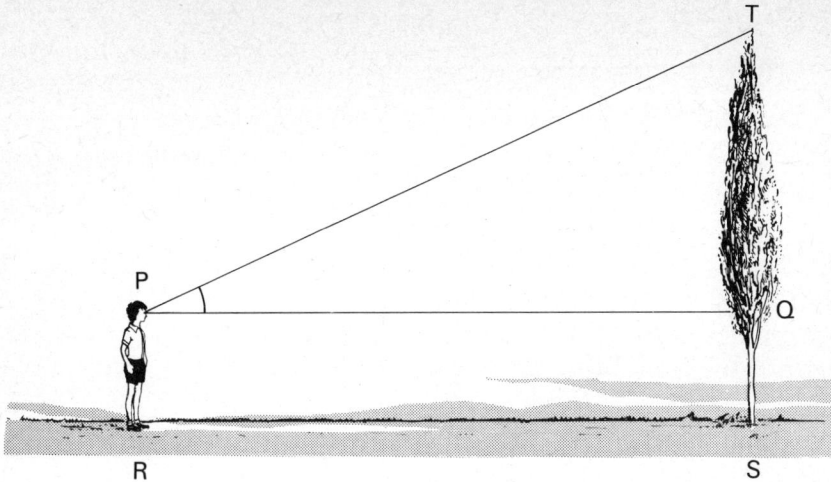

A boy is trying to find the height of a tree.
He decides that he can do this by measuring the length of RS and the marked angle at P.
With these two measurements he can make a scaled drawing of the triangle PQT, and so find the length of QT. Adding his own height to this gives the length of TS.

A simple instrument which can be used for measuring the angle at P is shown below. It is called a clinometer.

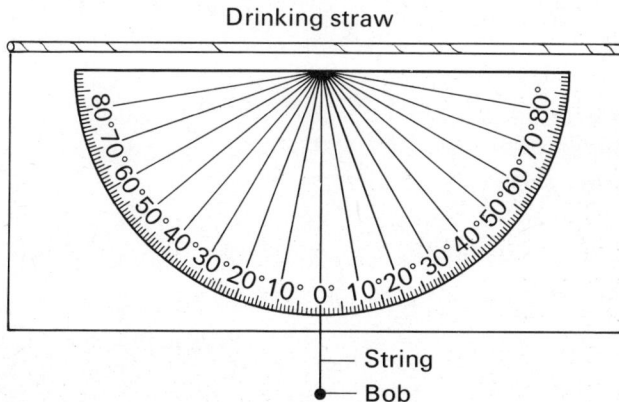

You can make a clinometer by using:
(a) a piece of thick card or hardboard
(b) a drinking straw
(c) a paper protractor, marked as above (W/s 68)
(d) a piece of fine string with a metal bob at one end.

A   Make a clinometer, as shown above.

A   With a friend, practise using your clinometer, by sighting the
     tops of trees, buildings, etc., through the drinking straw.

B   Start with the string over
     the 0° mark on the scale.
     Gradually tilt the straw
     until the string lies over
     the 20° mark.
     What angle does the straw
     now make with the horizontal?
     Repeat for other angles.

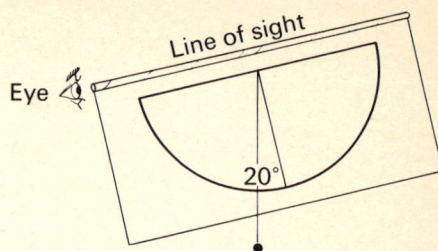

C   Look again at the drawing at the top of page 139.
     The boy found that the marked angle at P was 40°.
     He also found that the length of RS was 10 metres.
     Using these measurements make a scaled drawing of triangle PTQ.
     From this drawing find, to the nearest metre, the length of TQ.
     What do you think is the height of the tree, to the nearest metre?
     Compare your result with those of your friends.

D   Use the method described in C to find the heights of trees,
     buildings, etc. in the school grounds. For each object make sure
     that you can measure the distance from your viewing point to
     a point, on level ground, directly underneath the top of the
     object.

E   The drawing shows a block of
     flats on the other side of
     a busy road from a school.
     Suggest ways of finding its
     height without crossing the road.

F   Find the height of a building
     to which you cannot measure
     all the way along the ground.

The members of the Mathematics Club went to the seaside for a day. One group collected all kinds of information about birds, rocks, pebbles, and flowers. Another group found the height of the cliffs and of the flagpole at the coastguard station.

The third group decided to draw a map of the seashore, including three rocks which stood out from the sea, some distance from the land. They had brought with them a measuring tape, drawing paper, drawing pins, a wooden ruler, a ball of string, some poles, a piece of hardboard for a drawing board, and a stool.

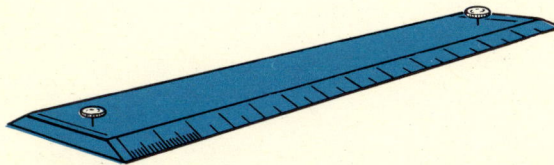

The group decided not to swim to each rock to measure their distance from the beach. They thought it would be simpler to use angles to fix the position of each rock. To help in this, they pushed two drawing pins into a ruler to make an alidade as shown above.

The group first put two poles P and Q in the sand, 100 m apart. They called PQ the *base line*. They then drew a line 20 cm long on a sheet of plain paper pinned to the hardboard. This line represented the distance between the poles.

A   What scale did the group use?

They then placed the hardboard and paper on the stool on the beach over the position of the pole P so that the 20-cm line on the paper pointed exactly to the pole Q. They did this by sighting along the drawing pins of the alidade placed along the line on the paper.

To obtain the angle between the base line QP and the Devil's Nose rock, they moved the alidade on the paper so that the drawing pins were exactly in line with the rock. They then drew a line on the paper through P towards this rock.

A   How would you draw on the paper the line at the correct angle
     from Q to the Devil's Nose rock?

B   Describe how to mark the position on the paper of the Devil's
     Nose rock.

The position of the Devil's Nose is
shown on the paper at the point D
where the line drawn from P and
the line drawn from Q meet. The
line from Q can be drawn only
by standing at pole Q and sighting
the rock from there.

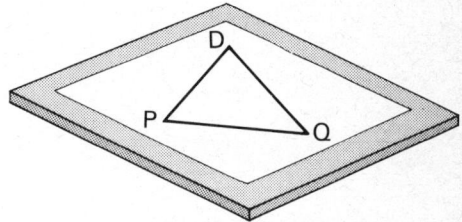

C   Explain how to find the position on the paper of the other two
     rocks, Mew Stone and Thatcher Rock.

To map the edge of the path between the
beach and the base of the cliff, the group
stretched a piece of string between the
two poles along the base line PQ. They
placed more poles at L and M and R along
this line. To find the position of R
they unwound more string in a straight
line with PQ.

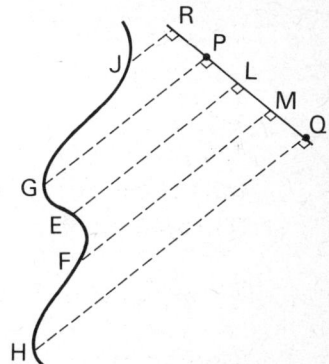

The group then measured the distance from L, M, P, Q, and R to
poles placed on the path at E, F, G, H, and J. They made sure that
the angles formed with the base line (as marked) were right angles as
near as could be judged by eye. Finally they measured the distances
PL, PM, and PR.

To complete the map, they drew these measurements to the same
scale as before on the paper. The path is offset from the base line.
The method is called an **offset survey**.

D   Use a base line and the methods described above to draw a map
     of your playing fields or some other suitable piece of ground.

Linear relations

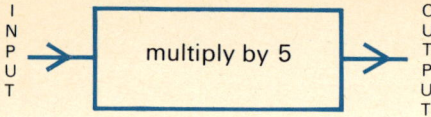

INPUT → | multiply by 5 | → OUTPUT

The machine above multiplies any number input by 5.
If 2 is input then the output is 10. This could be shown as (2, 10).
If 6 is input then the output is 30.

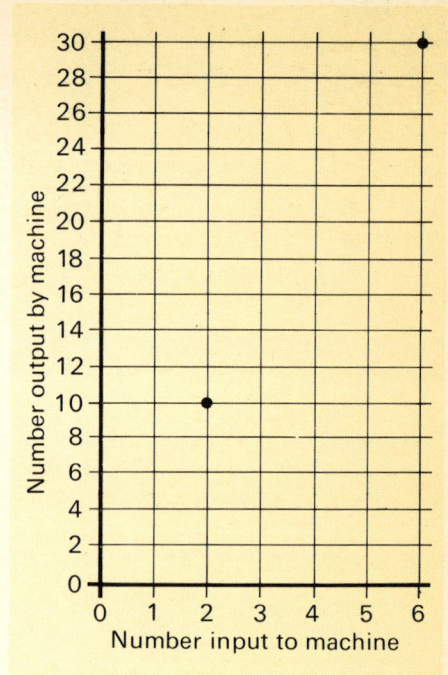

A   Input each of the numbers 1, 2, 3, 4, 5, and 6 to the *multiply by 5* machine. Find the number output. Show your results as ordered pairs.

B   Copy the graph above (W/s 69). Using the pairs you found in A, complete the graph.

Graph: y-axis labelled "Number output by machine" (0 to 30), x-axis labelled "Number input to machine" (0 to 6). Points shown at (2, 10) and (6, 30).

C   Input each of the numbers 1, 2, 3, 4, 5, and 6 to the machine shown on the right. Find the number output. Show your results as ordered pairs.

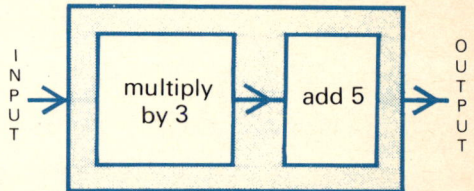

INPUT → | multiply by 3 | → | add 5 | → OUTPUT

D   Draw a graph to show the ordered pairs you found in C (W/s 69).

E   What can you say about the graphs you drew in B and D?

If the points on a graph lie on a straight line the graph is said to represent a **linear relation**.

F   Draw a graph (W/s 69) for the pairs:
1. (1, 4), (2, 5), (3, 6), (4, 7), (5, 8)
2. (1, 9), (2, 7), (3, 5), (4, 3), (5, 1)
3. (1, 2), (2, 5), (3, 10), (4, 17), (5, 26)

G   Which of your graphs in F represent linear relations?

A   Input each of the numbers 1, 2, 3, 4, 5, 6, 7, 8, 9, 10 to the machine
    below. Find the number output. Show your results as ordered pairs.

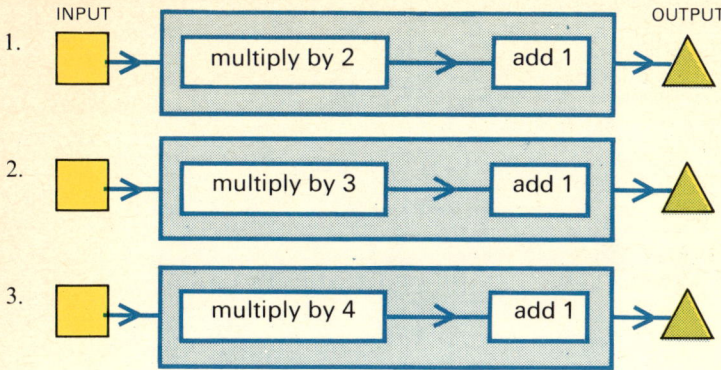

INPUT   OUTPUT

1. [ ] → [multiply by 2] → [add 1] → △

2. [ ] → [multiply by 3] → [add 1] → △

3. [ ] → [multiply by 4] → [add 1] → △

B   For each part of A write a □, △ statement.

C   Show the ordered pairs, in each part of A, on the same graph.
    You will need to be able to show the numbers from 0 to 10 on
    the □ axis, and numbers from 0 to 45 on the △ axis (W/s 70).
    Label each set of ordered pairs with its □, △ statement.

D   What do you notice about your three sets of ordered pairs in C?
    Does each set represent a linear relation?
    What can you say about the steepness of each line?
    Through which point on the △ axis does each line pass?

E   Look at the graph shown on
    the right.
    Write a □, △ statement for:
    1. the points on the blue
       line
    2. the points on the yellow
       line
    3. the points on the black
       line

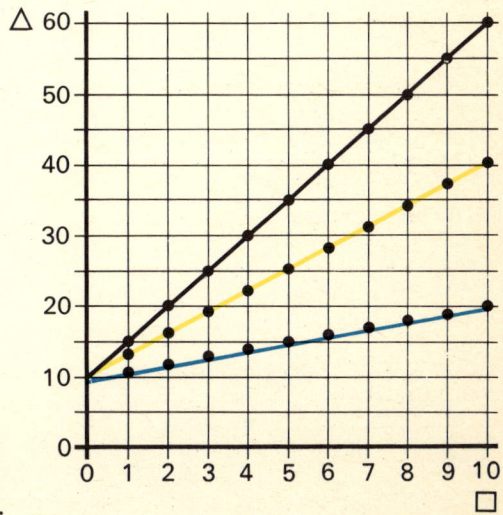

F   Through which point on the
    △ axis does each line pass?
    Could you tell this by looking
    at the □, △ statements?
    Check this with your graphs in C.

A   John is saving up to buy a tennis racket.
He has £3 already and is given 40p per
week as pocket money, which he does not spend.
Copy and complete the table below (W/s 71).

Start	Week 1	Week 2	Week 3	Week 4	Week 5	Week
£3	£3·40	£	£	£	£	

£8·60

B   Show the information in A as a graph (W/s 71).
One axis will show the different weeks.
The second axis will show the amount saved.

C   Look at the graph you have drawn in B.
Do the points lie on a straight line?
Where does this line cross the axis showing the amount saved?

D   Sally has to buy her own lunch every day, except Sunday.
Every four weeks, on a Monday, she is given £6. Her lunch costs
25p each day. Copy and complete the table below (W/s 72).

Start	day 1	day 2	day 3	day 4	day 5	day 6		day 27	day 28
£6	£5·75								

E   Show the information in D as a graph (W/s 72). One axis will show
the day. The second axis will show the amount Sally has left.
Does this graph represent a linear relation?

F   The table below shows the speed a car
reaches from a standing start, after
a given number of seconds.

Time in s	2	4	6	8	10	12
Speed in km/h	4	16	30	44	56	60

Show this information as a graph (W/s 73).
One axis will show the time in seconds.
The second axis will show the speed in km/h.
Describe the shape of this graph.
Try to explain why it has this shape.

A   Input each of the numbers ⁻5, ⁻4, ⁻3, ⁻2, ⁻1, 0, 1, 2, 3, 4, 5 to the
    machine below. Find the number output and show your results as
    ordered pairs.

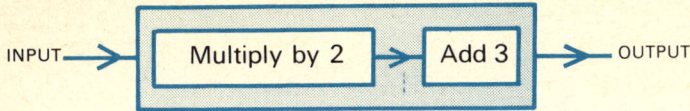

    INPUT ⟶ [Multiply by 2] ⟶ [Add 3] ⟶ OUTPUT

B   Write down the □, △ statement for A, and show the ordered
    pairs on a graph (W/s 73). You will need to be able to show the
    numbers ⁻5 to 5 on the □ axis and ⁻7 to 13 on the △ axis.

C   The points in B lie on a line. At what point does this line cross
    the △ axis? What can you say about the steepness of the line?

    Using letters, if x stands for the number input and y for the number
    output, then for the machine above $y = x \times 2 + 3$ or $y = 2x + 3$.

D   Draw a diagram, as in A, to show the machine for which:
    1. $y = 5x + 2$ 2. $y = 3x + 7$ 3. $y = 2x - 5$

E   For each part of D complete the table (W/s 74).

x	⁻5	⁻4	⁻3	⁻2	⁻1	0	1	2	3	4	5
y											

F   Show on the same graph each set of ordered pairs (x, y) in E
    (W/s 74).

G   Each of your graphs in F represents a linear relation.
    Can you tell by looking at the equations in D:
    1. which is the steepest line?
    2. which is the least steep line?
    3. at what point each line crosses the y axis?

H   Using x as a member of {0, 1, 2, 3, 4, 5, 6, 7, 8} draw on the same
    sheet of graph paper (W/s 75):
    1. $y = 2x + 1, y = 2x + 3, y = 2x + 4, y = 2x + 5$
    2. $y = x + 5, y = 2x + 5, y = 3x + 5, y = 4x + 5$

I   Sketch (W/s 76), without having to find all the ordered pairs, the
    graph for: 1. $y = 2x + 2$ 2. $y = 2x + 7$ 3. $y = 4x + 3$
    Assume that x can be any number from 0 to 10.

Mr. Graham is trying to decide whether it is cheaper to run his central heating on oil or on gas.

He is told that for oil he will have to pay £10 per year plus 12p for each litre of oil used.

In the case of gas he has to pay £16 per year plus 9p for each litre of gas used.

A   How much would Mr. Graham have to pay in a year if he used:
1. 200 litres of oil?   2. 300 litres of gas?

B   Use the flow chart below to find the ordered pairs:
(number of litres of oil used, cost of oil per year)
for 100, 200, 300, 400, 500, 600, 700 litres of oil.

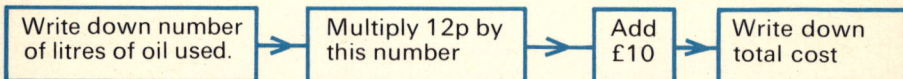

| Write down number of litres of oil used. | → | Multiply 12p by this number | → | Add £10 | → | Write down total cost |

C   Show the ordered pairs in B on a graph (W/s 77).
What can you say about the set of points on your graph?

D   Make up a flow chart like the one above, to help you to find the cost of using a given number of litres of gas.
Use your flow chart to find the cost of using 100, 200, 300, 400, 500, 600, 700 litres of gas.
Show this information on your graph for C (W/s 77).

F   For how many litres would the cost of gas be the same as oil?

The graph on the right shows $y = 3x + 2$, where x is any number from 0 to 8.

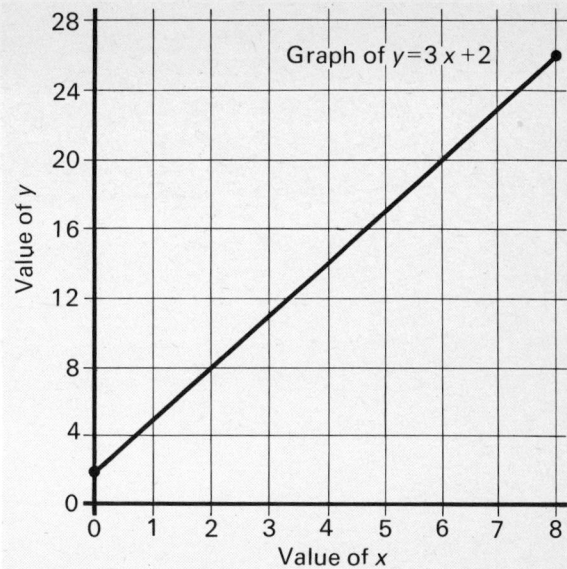

Graph of $y = 3x + 2$

Value of y

Value of x

A   Write down two or three ordered pairs which make the equation $y = 3x + 2$ true.

B   Write down two or three ordered pairs which make the equation $y = 2x + 8$ true.

C   Copy the graph on the right (W/s 77). Show on it, as well, $y = 2x + 8$.

D   Use your graph in C to find an ordered pair which makes both the equations, $y = 3x + 2$ and $y = 2x + 8$, true at the same time.

E   Draw a graph (W/s 78) to show $y = 4x + 1$ and $y = 2x + 11$. Use your graph to find the ordered pair which makes both the equations true at the same time.

F   The ordered pairs (1, 5), (2, 9), (3, 13), (4, 17), (5, 21) make the equation $y = 4x + 1$ true.
The ordered pairs (1, 13), (2, 15), (3, 17), (4, 19), (5, 21) make the equation $y = 2x + 11$ true.
What is the intersection of these two sets of ordered pairs?
At what point do your two lines in E intersect? Comment.

H   A man is m years old. His son is s years old. They have a total age of 46 years. Write down a statement showing the relationship between m and s.
The man is four times as old as his son plus one year.
Write down a second statement showing another relationship between m and s.
Find the ages of the man and his son.
Check your answer by drawing a graph (W/s 78) for the two linear relations and finding their intersection.

Reading graphs

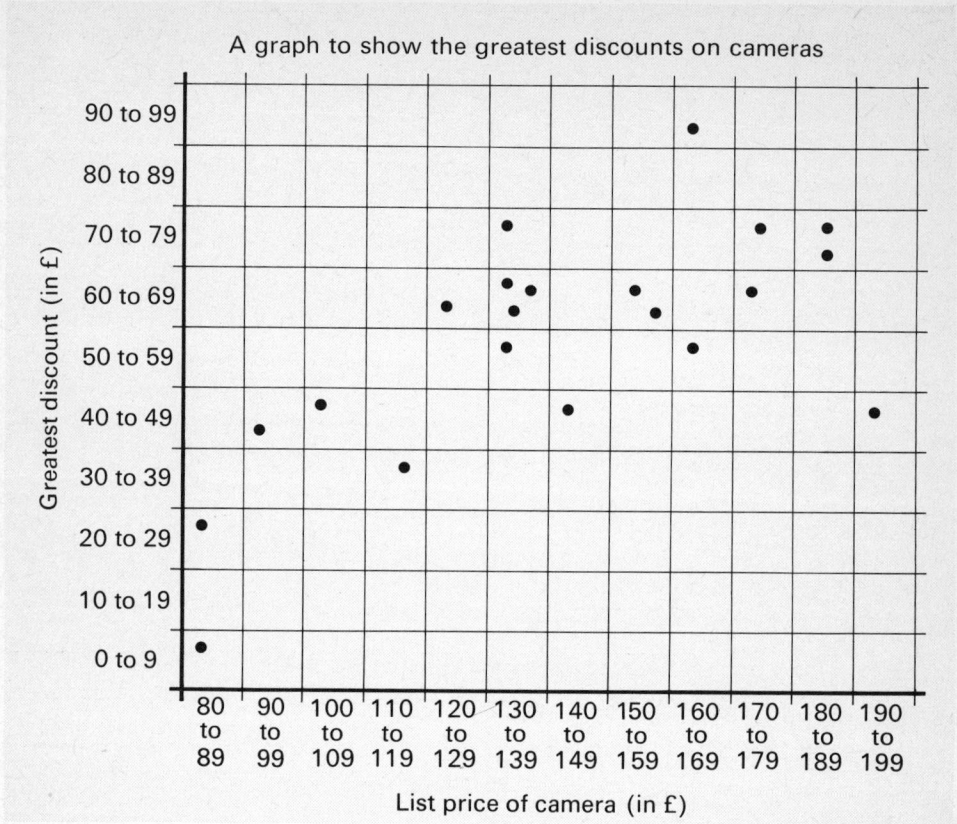

A graph to show the greatest discounts on cameras

In the May 1973 issue of *Which?* details were given of the list prices of a number of single lens reflex cameras. It also gave the lowest price at which it had been found each camera could be bought. The above graph is based on this information.

A   Write down all that you can find out from the graph.

B   Do you think it is true to say that *the dearer the camera the more discount you can get*?

C   Do you think it would have been better to have shown the saving as a percentage of the list price?
Try doing this from the information given in the above graph.
Show your results on a graph (W/s 79). What do you notice?

A graph to show the number of pages and the costs of a set of fiction books

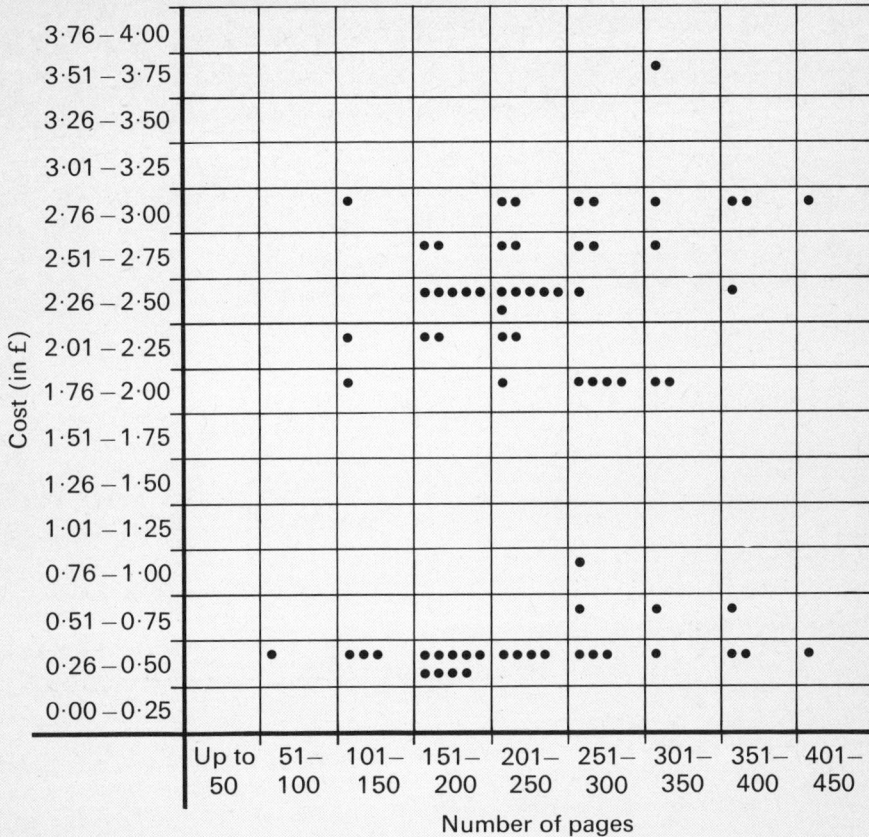

Cost (in £)	Up to 50	51–100	101–150	151–200	201–250	251–300	301–350	351–400	401–450
3·76 – 4·00									
3·51 – 3·75							•		
3·26 – 3·50									
3·01 – 3·25									
2·76 – 3·00			•		••	••	•	••	•
2·51 – 2·75				••	••	••	•		
2·26 – 2·50				•••••	•••••• •	•		•	
2·01 – 2·25			•	••	••				
1·76 – 2·00			•		•	••••	••		
1·51 – 1·75									
1·26 – 1·50									
1·01 – 1·25									
0·76 – 1·00						•			
0·51 – 0·75						•	•	•	
0·26 – 0·50	•		•••	••••• ••••	••••	•••	•	••	•
0·00 – 0·25									

Number of pages

A   Look at the graph. Write down all you can find out from it.

B   There are two distinct subsets of points in the graph—a lower set and an upper set. Can you suggest reasons for this?

C   The above graph was based on 1973 prices. Obtain information about the current prices of a sample of fiction books. Show your results on a graph (W/s 80). Compare your graph with the one above. What do you notice?

D   Do you think it is true to say that *the more pages there are in a book, the higher is its price*?

A graph to show the distance travelled on a litre of petrol for cars of various engine size

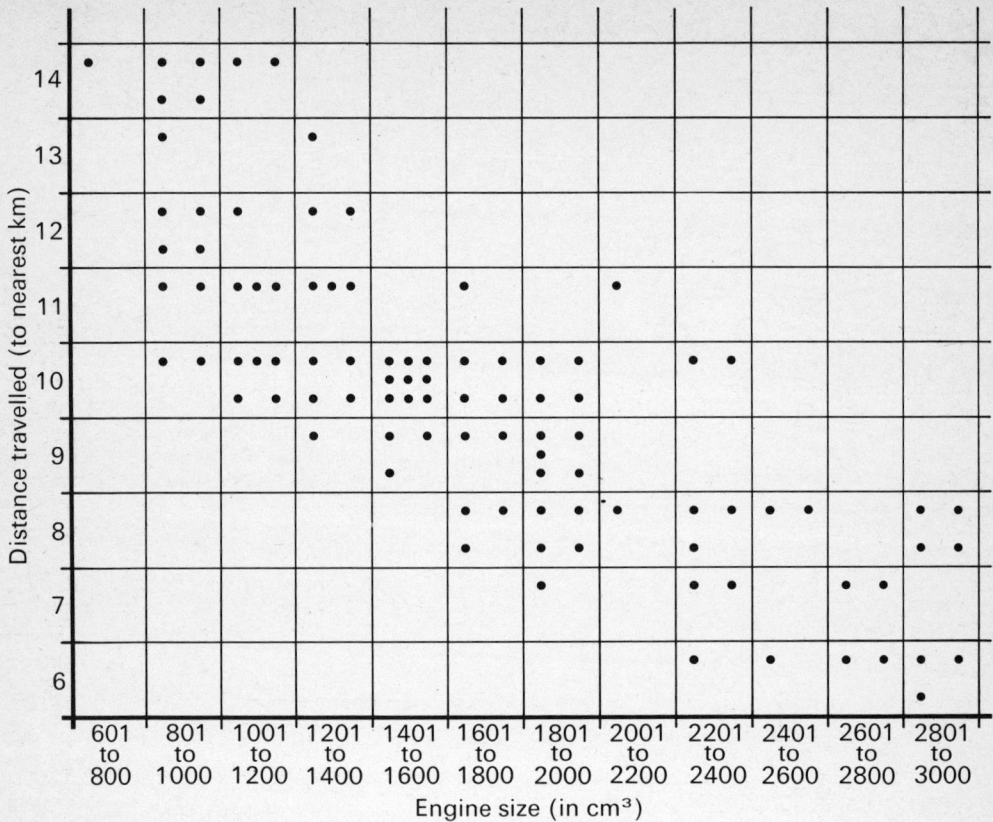

| | 601 to 800 | 801 to 1000 | 1001 to 1200 | 1201 to 1400 | 1401 to 1600 | 1601 to 1800 | 1801 to 2000 | 2001 to 2200 | 2201 to 2400 | 2401 to 2600 | 2601 to 2800 | 2801 to 3000 |

Engine size (in cm³)

A   The graph shows the distance travelled on a litre of petrol for cars of various engine sizes.
Write down all you can find out from the graph.

B   Do you think it is true to say that *the larger the engine, the shorter the distance you travel on a litre of petrol?*

C   If you agree that there is some relationship between engine size and distance travelled on a litre of petrol, can you suggest a reason for it?

D   Try drawing a graph (W/s 81) to show the relationship between the engine size of a car and its maximum speed.

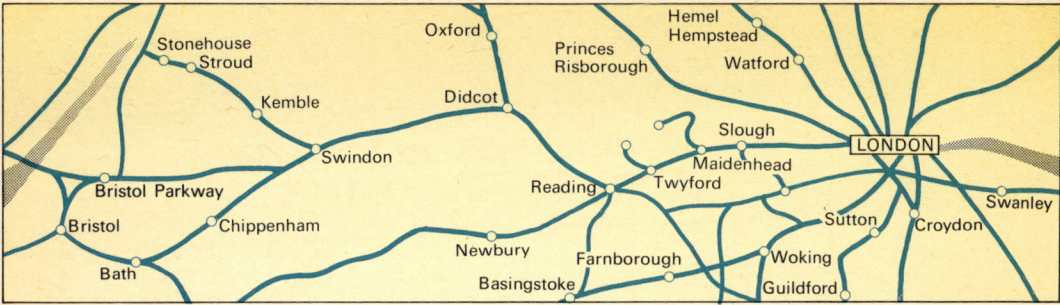

A graph to show how many trains from Paddington stop at various stations

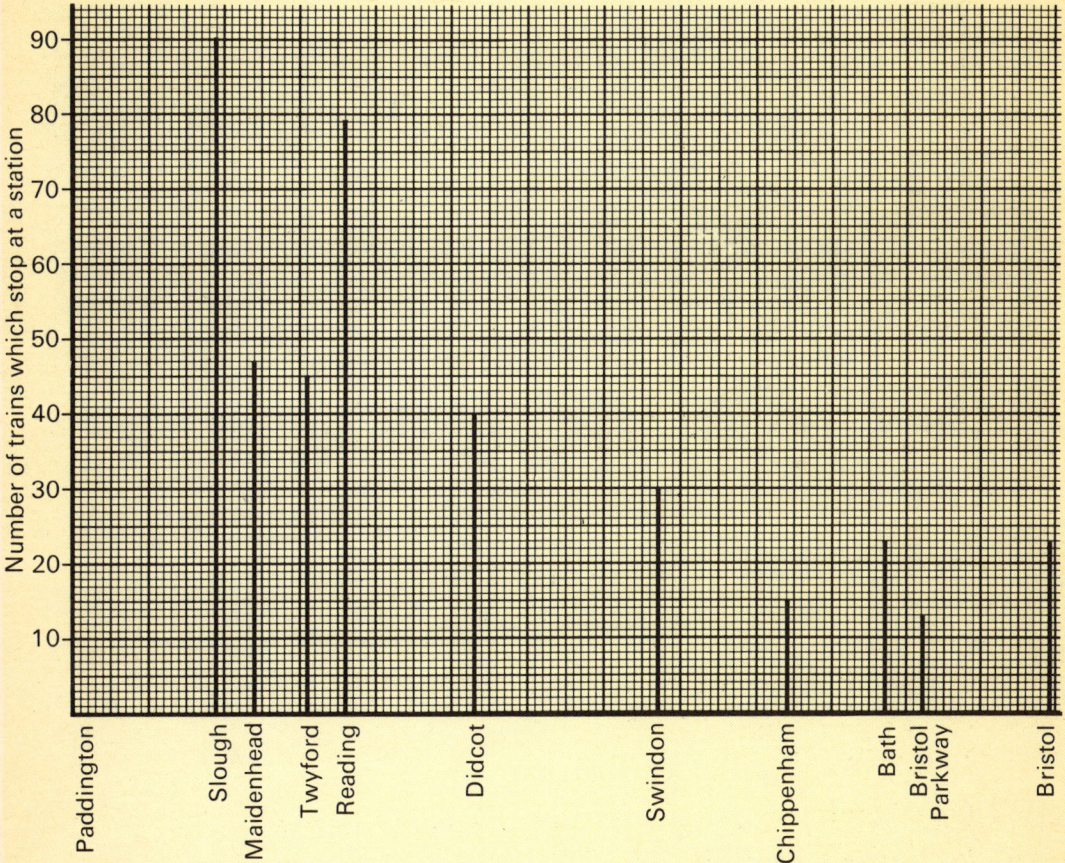

A   Look at the map of the rail routes from London (Paddington) to Bristol and the graph below it. Write all you find out from them.

B   Suggest reasons why: 1. Slough has most trains   2. Didcot has more than Swindon   3. Bath and Bristol have the same number 4. Chippenham has less than Bath   5. Reading has more than Maidenhead and Twyford but less than Slough.

A graph to show my
marks in mathematics

A graph to show my
marks in mathematics

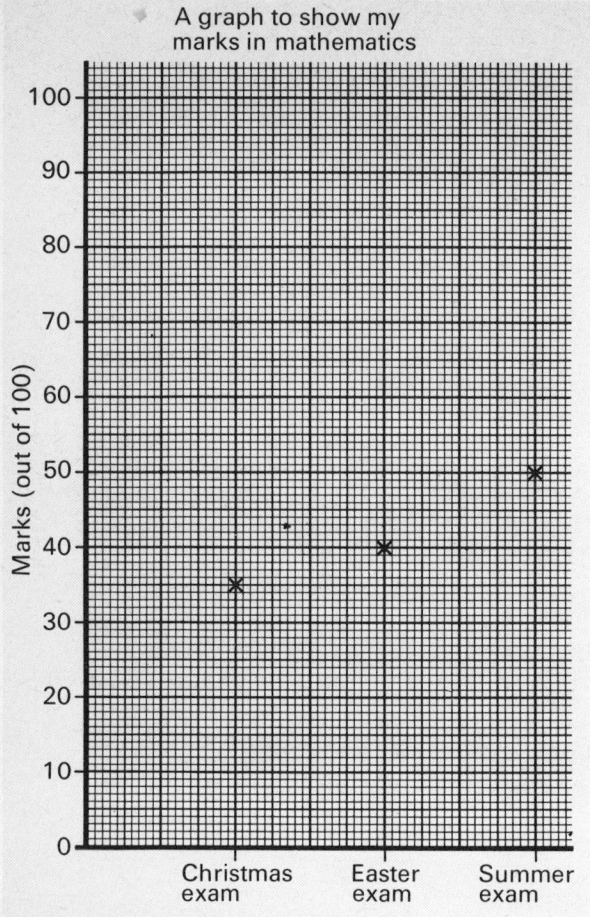

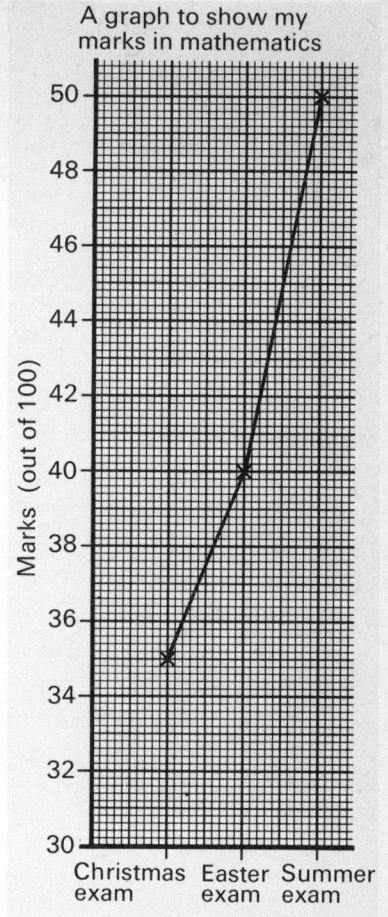

Peter had not done very well in his mathematics examination
at the end of the first term. At the end of each of the two next
terms he did rather better. He decided to draw a graph to show
his parents how he had improved.

First, he drew the graph shown on the left above. He did not
think, however, that this looked very exciting. So, after a little
thought, he redrew it to get the graph shown on the right. He
thought this showed a much better improvement.

A   What do you think of the second graph? Does it show the marks
correctly? Does it give a true impression of his progress?

B   How do you think Peter could have shown his marks if they had
been 70, 65, and 60 in this order?

Radioactive decay

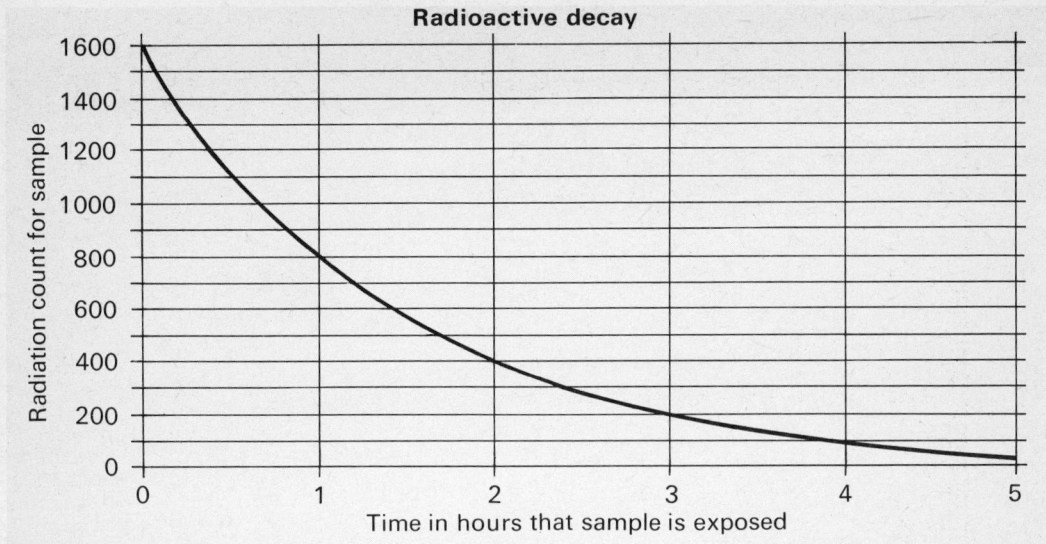

A   The graph above shows how the radiation count for a particular
     radioactive sample changes with its age.
     The initial radiation count is 1600. After what length of time
     will the radiation count be halved? After what further length of time
     will it be halved again? What do you notice? What will be the
     radiation count after the same amount of time again?
     After how many hours will the radiation count be less than 10?

THE GO-AHEAD PARTY

SEE HOW MEMBERSHIP IS INCREASING

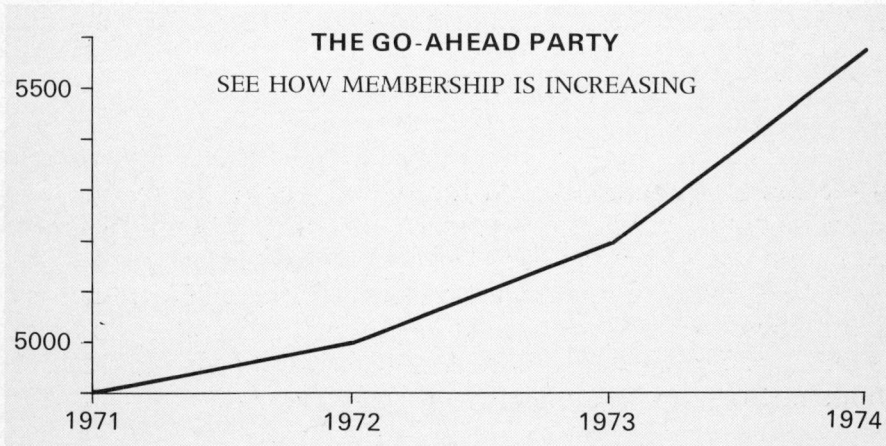

B   Look at the above graph. What can you say about how the
     membership has increased in each of the last three years?
     If membership continues to increase in the same way, in what year
     will it reach 10 000? Do you think that this is an honest graph?

Review 3A

A   Estimate and then calculate the circumference of the circle. Use: 1. $\pi = 3 \cdot 14$ 2. $\pi = 3\frac{1}{7}$

B   Write, using an index:
1. $3 \times 3 \times 3 \times 3 \times 3$ 2. $7 \times 7 \times 7 \times 7$

C   Find the difference between 2^5 and 5^2.

D   Copy the green shape on squared paper (W/s 81). Draw the image of the shape after it has been moved by the translation:

1. $\begin{pmatrix} 2 \\ 1 \end{pmatrix}$ 2. $\begin{pmatrix} 0 \\ 2 \end{pmatrix}$ 3. $\begin{pmatrix} -1 \\ 1 \end{pmatrix}$

E   1. Write down the co-ordinates of the vertices of the blue shape. Give the letter and the co-ordinates for each vertex.

2. Write down the co-ordinates of P when the shape is given a clockwise rotation about the origin of: (a) 90° (b) 180° (c) 270°.

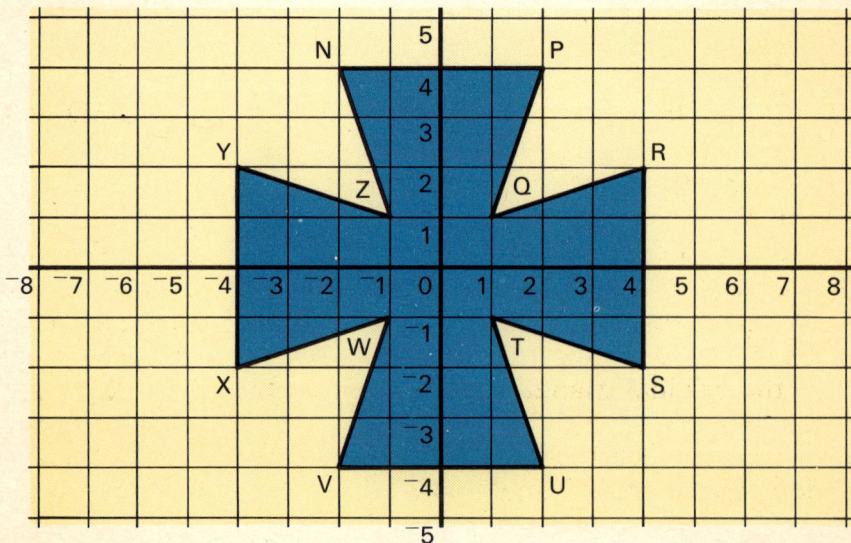

F   The rectangle is drawn to a scale
     of 1 : 20.
     1. Find the length and width
        of the rectangle represented
        by the drawing.
     2. Find the length of a diagonal
        of the original rectangle.
     3. By what number must you multiply the area of the scaled
        drawing to get the area of the original rectangle?

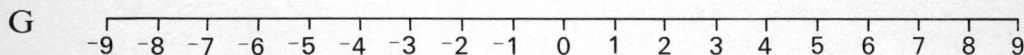

G
     ⌐─┬──┬──┬──┬──┬──┬──┬──┬──┬──┬──┬──┬──┬──┬──┬──┬──┬──┬─┐
     -9 -8 -7 -6 -5 -4 -3 -2 -1   0   1   2   3   4   5   6   7   8   9

     On a number line, like that above (W/s 82) mark the set of points
     for which the statement is true:
     1. $\square > 3$
     2. $\square < {}^{-}2$
     3. $\square = 5$
     4. $\square > 0$ and $\square < 4$
     5. $\square < 5$ and $\square > {}^{-}2$
     6. $\square < 3$ and $\square > 4$
     7. $\square + 2 > 1$
     8. $\square - 2 < {}^{-}1$

H   If $\square$ and $\triangle$ are members of {integers from 0 to 10} write down
     the solution set for $\square + \triangle = 9$.
     Show the ordered pairs as points on a graph (W/s 83).
     For {integers from 0 to 10} write down the solution set for
     $\triangle - \square = 3$. Show this set on the same graph as before.
     Write down the solution set for $\square + \triangle = 9$ and $\triangle - \square = 3$.

I   (1, 2), (1, 4), (4, 2) are the vertices of
     a triangle. Using each vertex in turn
     work through the program on the right.
     The numbers output in ⑦ and ⑧ are
     the co-ordinates for the image of the
     vertex you used.
     Show the original triangle and its
     image on the same graph (W/s 83).
     What transformation would you use
     to move the original triangle onto its
     image?

     ①   3 → **R**
     ②   2 → **U**
     ③ Enter the first
        co-ordinate in box   **X**
     ④ Enter the second
        co-ordinate in box   **Y**
     ⑤   **X + R → X**
     ⑥   **Y + U → Y**
     ⑦   OUTPUT   **X**
     ⑧   OUTPUT   **Y**
     ⑨   STOP

Review 3B

A | Using $\pi \simeq 3\cdot14$, find the diameter of a circle whose circumference is:
1. 27 cm   2. 83·4 cm   3. 5 m   4. 19·27 m

B | Find the value of: 1. $1\cdot2^2$ 2. $4\cdot1^2$ 3. $1\cdot3^3$ 4. $0\cdot2^4$

C | Try to find: 1. $\sqrt{225}$ 2. $\sqrt{400}$ 3. $\sqrt{900}$ 4. $\sqrt{625}$

D | Write down the vectors which have been used to translate a point from:
1. P to Q
2. Q to R
3. R to S
4. S to T
What single vector would translate a point from P to T?
What single vector would translate a point from T to P?

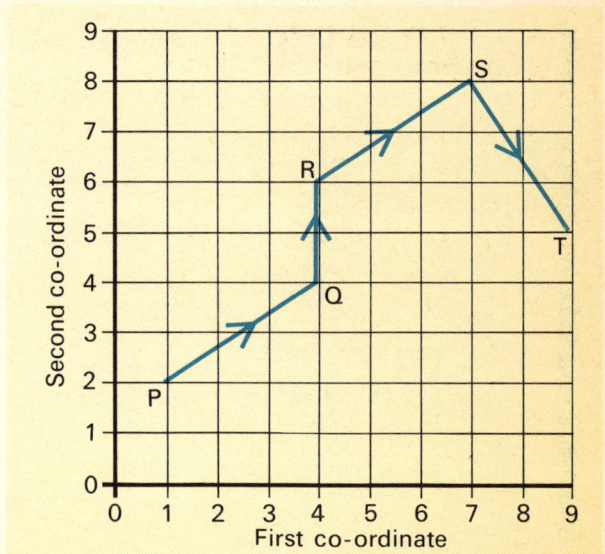

E | Write down the answer:
1. $3 + {}^-2$ 2. $4 + {}^-5$ 3. ${}^-4 + {}^-7$ 4. $5 - 3$ 5. $3 - 7$ 6. $4 - {}^-3$
7. $12 - {}^-14$ 8. ${}^-15 - {}^-20$ 9. ${}^-42 - {}^-20$ 10. $0 - 7$ 11. $0 - {}^-15$

F | Find the value of:
1. $(2 + 3) \times 5$ 2. $(8 + {}^-5) \times 4$ 3. $({}^-3 - {}^-6) \times 8$ 4. $(7 - {}^-1) \times {}^-9$

G | The rough sketch on the right shows the measurements recorded whilst finding the height of a tree.
Find the height of the tree by using the measurements to make an accurate drawing.

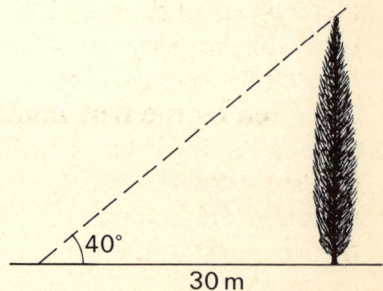

40°
30 m

H A blue die and a green die
are thrown at the same time.

Using *B* for the score
on the blue die,
G for the score
on the green die,
all the possible ordered
pairs (*B*, *G*) are shown on
the graph on the right.

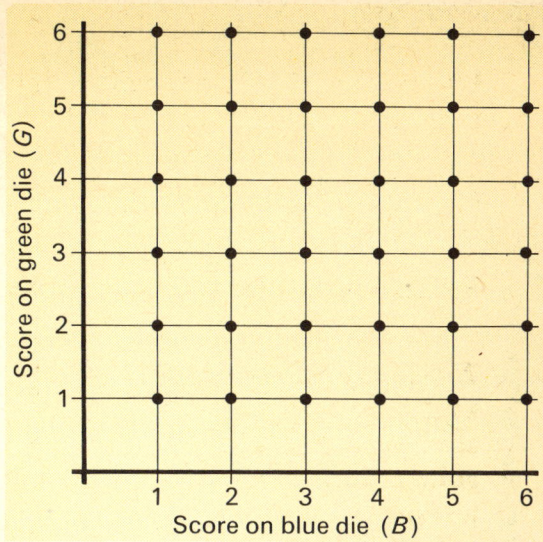

Use the graph to find the set
of ordered pairs for which:
1. $B > 3$ and $G < 3$
2. $B + G = 8$
3. $G = B + 2$
4. $G = B \times 2$
5. $B + G < 8$ and $B + G > 4$
6. $G = B + 2$ and $G = B \times 2$

Score on green die (*G*)

Score on blue die (*B*)

INPUT → Multiply by 2 → Subtract 3 → OUTPUT

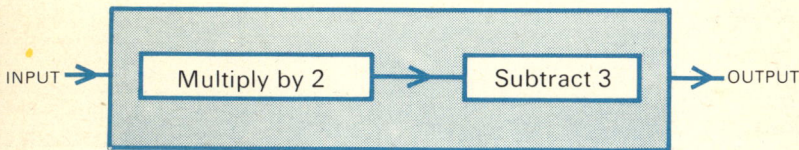

I Input each member of {⁻4, ⁻3, ⁻2, ⁻1, 0, 1, 2, 3, 4} into the machine
shown above. Find the number output.
Show your results as ordered pairs and as points on a graph (W/s 84).

J Work through the program on the right.
Draw a triangle whose edges, in cm, are
given by the numbers in **P**, **Q**, and **R**.
What can you say about this triangle?
Now put 5 in **P** and 12 in **Q**, and work
through the program again.
Draw the triangle whose edges, in cm,
are given by the new numbers in **P**, **Q**,
and **R**. What do you notice?
Choose other pairs of numbers to put
in **P** and **Q**. For each draw a triangle,
as before. What do you notice?

①
②   3 → **P**
③   4 → **Q**
④   **P** × **P** → **W**
⑤   **Q** × **Q** → **V**
⑥   **W** + **V** → **S**
⑦ $\sqrt{S}$ → **R**
⑧   OUTPUT   **R**
            STOP

Review 3C

A   Try to find a pair of numbers, each less than five, for which
$$a^b = b^a$$

B   Write in a shorter form:
1. $(p \times p \times p) + (p \times p \times p) + (p \times p \times p) + (p \times p \times p)$
2. $l^2 + 2m^2 + 3l^2 + 2m^2$
3. $4q^4 + 4q^4$

C   A point P with co-ordinates (x, y) is translated, using $\binom{a}{b}$ followed by $\binom{c}{d}$. What are the co-ordinates of the final image point?
What single translation could replace the two translations?

D x and y are both members of {whole numbers from 1 to 20}
X is the solution set for $x \times 3 > 24$
Y is the solution set for $y + 8 < 24$
Find: 1. $X \cap Y$ 2. $X \cup Y$ 3. $X \setminus Y$ 4. $Y \setminus X$

E x is a member of {integers from $^-2$ to 10}
y is a member of {integers from 0 to 24}
Show on a graph (W/s 85) the set of ordered pairs which make the statement $y = 2x + 3$ true.
On the same graph show the set of ordered pairs which make the statement $y = 12 - x$ true.
Use your graph to find the ordered pair which makes both equations true.

F   The sum of two numbers is 11. Twice the smaller number is one more than the other number. If the numbers are represented by x and y write down two equations connecting them. Find the two numbers.

G   Work through the program.
Record the numbers output in
⑤ and ⑥ as co-ordinates (x, y).
Show these pairs as points on a
graph (W/s 85).
What do you notice about this set
of points?
Find the equation representing
this set of points.

①	$1 \rightarrow$ **A**
②	$0 \rightarrow$ **X**
③	$\mathbf{X} + \mathbf{X} \rightarrow$ **W**
④	$\mathbf{W} + \mathbf{A} \rightarrow$ **Y**
⑤	OUTPUT **X**
⑥	OUTPUT **Y**
⑦	$\mathbf{X} + \mathbf{A} \rightarrow$ **X**
⑧	IF $\mathbf{X} < 10$ GO TO ③
⑨	STOP